S0-ADM-751

FRACTIONS

Globe
Fearon

Upper Saddle River,
New Jersey

Executive Editor: Barbara Levadi
Editors: Bernice Golden, Lynn Kloss, Robert McIlwaine, Kirsten Richert
Production Manager: Penny Gibson
Production Editor: Walt Niedner
Interior Design: The Wheetley Company
Electronic Page Production: Curriculum Concepts
Cover Design: Pat Smythe

Printed in the United States of America   4 5 6 7 8 9 10 04 03 02 01 00

ISBN 0-8359-1547-6

CONTENTS

TO THE STUDENT

Access to Math is a series of 15 books designed to help you learn new skills and practice these skills in mathematics. You'll learn the steps necessary to solve a range of mathematical problems.

LESSONS HAVE THE FOLLOWING FEATURES:

❖ Lessons are easy to use. Many begin with a sample problem from a real-life experience. After the sample problem is introduced, you are taught step-by-step how to find the answer. Examples show you how to use your skills.

❖ The *Guided Practice* section demonstrates how to solve a problem similar to the sample problem. Answers are given in the first part of the problem to help you find the final answer.

❖ The *Exercises* section gives you the opportunity to practice the skill presented in the lesson.

❖ The *Application* section applies the math skill in a practical or real-life situation. You will learn how to put your knowledge into action by using manipulatives and calculators, and by working problems through with a partner or a group.

Each book ends with *Cumulative Reviews*. These reviews will help you determine if you have learned the skills in the previous lessons. The *Selected Answers* section at the end of each book lists answers to the odd-numbered exercises. Use the answers to check your work.

Working carefully through the exercises in this book will help you understand and appreciate math in your daily life. You'll also gain more confidence in your math skills.

Vocabulary

factors: numbers that are multiplied to find a product

prime number: a number greater than 1 that has exactly two factors, itself and 1

prime factor: a number expressed as the product of prime numbers

FACTORS

Problem 1: A store clerk arranges 36 CDs in rows so that each row has the same number of CDs. How many arrangements are possible? One way to find out is to look at the factors of 36.

$$18 \times 2 = 36 \qquad 6 \times 3 \times 2 = 36$$
$$\text{factors} \qquad\qquad \text{factors}$$

Use multiplication sentences to find all of the factors of a whole number.

$$1 \times 36 = 36$$
$$2 \times 18 = 36$$
$$3 \times 12 = 36$$
$$4 \times 9 = 36$$
$$6 \times 6 = 36$$

Factors of 36: 1, 2, 3, 4, 6, 9, 12, 18, 36

There are 9 arrangements of equal rows that can be made from 36 CDs: 1 row of 36 CDs, 2 rows of 18, 3 rows of 12, ... , 36 rows of 1 CD.

Problem 2: 11 and 13 are prime numbers. 11 has exactly two factors, 11 and 1. 13 also has exactly two factors.

$$1 \times 11 = 11 \qquad \text{Factors of 11: 1, 11}$$
$$1 \times 13 = 13 \qquad \text{Factors of 13: 1, 13}$$

Prime numbers: 2, 3, 5, 7, 11, 13, 17, 19,...

To find the **prime factors** of a whole number, divide by prime numbers, beginning with 2. Keep dividing by 2 until it no longer goes into the number evenly; then go on to 3.

$$\begin{array}{r} 2\overline{)36} \\ 2\overline{)18} \\ 3\overline{)9} \\ 3 \end{array}$$

Divide by 2, then 3, and so on.

Stop when you get a prime number.

Prime factorization of 36: $2 \times 2 \times 3 \times 3$

1. List all the factors of 24.

$1 \times 24 = 24$ _____ $\times$ _____ = _____

$2 \times 12 =$ _____ _____ $\times$ _____ = _____

Factors of 24: _____

2. Copy and complete to find the prime factors.

a. 2 | 20
      2 | 10
          | ? $2 \times 2 \times ?$

b. 2 | 42
      ? | 21
          | ? $2 \times ? \times ?$

Exercises

List all the factors of each number.

3. 20   **4.** 35   **5.** 18

_____   _____   _____

6. 23   **7.** 60   **8.** 72

_____   _____   _____

Find the prime factors of each number.

9. 32   **10.** 54   **11.** 66

_____   _____   _____

12. 72   **13.** 95   **14.** 120

_____   _____   _____

Application

15. A friend of yours has missed this lesson. Write an explanation to tell your friend how to find the prime factors of 30.

_____

_____

LEAST COMMON MULTIPLE

Vocabulary

multiple of a number: the product of two numbers is a multiple of each of the numbers

least common multiple: the smallest nonzero common multiple of two or more numbers

One company's ships come to port every 3 days; another company's arrive every 4 days; a third company's come every 8 days. How many days will pass before the ships from each company come to port on the same day? You can find out by using multiples.

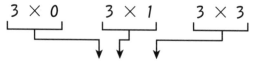

Multiples of 3: 0, 3, 6, 9, 12, 15, 18, 21, 24, . . .

Multiples of 4: 0, 4, 8, 12, 16, 20, 24, 28, 32, . . .

Multiples of 8: 0, 8, 16, 24, 32, 40, 48, 56, 64, . . .

The common multiples of two (or more) numbers are the nonzero numbers found in both (or all) lists.

Common multiples of 3 and 4: 12, 24, 36, . . .

Common multiples of 4 and 8: 8, 16, 24, . . .

Common multiples of 3, 4, and 8: 24, 48, 72, . . .

Find the least common multiple (LCM) of 3, 4 and 8 to find the first day all the ships return together.

The LCM of 3, 4, and 8 is 24. So, the ships will all be at port again in 24 days.

Guided Practice

1. Find common multiples of 2 and 3.

a. List nonzero multiples of 2 and 3 in order.

2: 2, 4, _____

3: 3, _____

b. Compare lists. Write three common multiples

of 2 and 3. _____

2. Find the LCM of 12 and 15.

a. List nonzero multiples of 12 and 15 in order.

12: 12, 24, 36, _____

15: 15, _____

b. Choose the least number that is in both lists.

LCM of 12 and 15 = _____

Exercises

List the first ten nonzero multiples of each number.

3. 2

4. 5

5. 6

6. 11

7. 20

8. 60

Find the LCM for each set of numbers.

9. 3, 6

10. 5, 9

11. 10, 20

12. 2, 3, 4

13. 4, 12, 24

14. 10, 15, 45

Application

 COOPERATIVE LEARNING

In each column, find the LCM of each pair of numbers.

15. Column A

2 and 3 _____

4 and 5 _____

6 and 7 _____

8 and 9 _____

Column B

2 and 14 _____

3 and 12 _____

5 and 10 _____

7 and 21 _____

Column C

2 and 3 _____

5 and 7 _____

11 and 13 _____

17 and 19 _____

16. Describe any shortcuts you have developed to find LCMs.

_____

_____

_____

FRACTIONS AND MIXED NUMBERS

fraction: a number in the form of $\frac{a}{b}$. Both a and b are whole numbers and b cannot be equal to 0.

mixed number: the sum of a whole number and a fraction

How could you express "3 tools out of a total of 5 tools" or "1 part out of a total of 3 equal parts" more simply?

A **fraction** names part of a whole. The whole can be a group, an object like a container, or an area.

3 of the 5 tools are wrenches.
$\frac{3}{5}$ of the tools are wrenches.

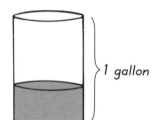

1 of the 3 equal parts is filled.
$\frac{1}{3}$ of the container is filled.

A **fraction** is made up of:

numerator: 3 ⟵ *number of parts being considered*
———
denominator: 5 ⟵ *number of equal parts in the whole*

A **mixed number** is made up of a whole number and a fraction.

whole number: 2 | mixed number: $2\frac{1}{2}$

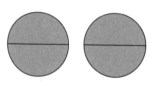 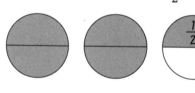

$2\frac{1}{2}$ means $2 + \frac{1}{2}$

Guided Practice

1. Write the fraction that expresses each amount:

   a. half of the fifty states: ____$\frac{25}{50}$____

   b. 7 parts out of a total of 20 equal parts : _____

2. Write a mixed number that names the shaded parts.

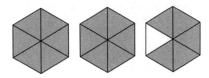

  a. Write the whole number part:

  _____

  b. Write the fraction part: _____

  c. Add the parts to get a mixed number: _____

Exercises

Write a fraction that names the shaded part of the area or group:

3.

_____

4.

_____

5.

_____

Write a mixed number that names the shaded parts:

6.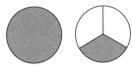

_____

7.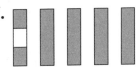

_____

Write a fraction for the situation:

8. Of the 24 people in the group, 9 are left-handed. What part of the group is left-handed? What part is not left-handed? _____

Application

Use the quilt to write a mixed number for the parts named.

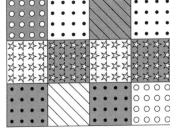

9. 3 whole quilts and any 4 squares: _____

10. 5 whole quilts, with one of them missing its striped squares: _____

COOPERATIVE
LEARNING

11. Describe a mixed number using quilts and squares from a quilt. Ask a partner to write the mixed number you describe:

🖊 _____

EQUIVALENT FRACTIONS

Reminder

$\frac{2}{2} = 1$, $\frac{3}{3} = 1$, $\frac{4}{4} = 1$, and so on.

Phil works after school as a data entry operator. Last week, he spent $\frac{1}{2}$ of his time creating new files, $\frac{1}{4}$ of his time making corrections, and $\frac{2}{8}$ of his time formatting disks. Did he spend the same amount of time on any of his tasks? Which ones?

Fractions that name the same part of a group, area, or object are called **equivalent equations.**

1								
$\frac{1}{2}$				$\frac{1}{2}$				$\frac{1}{2}$
$\frac{1}{4}$		$\frac{1}{4}$		$\frac{1}{4}$		$\frac{1}{4}$		$\frac{2}{4}$
$\frac{1}{8}$	$\frac{1}{8}$	$\frac{1}{8}$	$\frac{1}{8}$	$\frac{1}{8}$	$\frac{1}{8}$	$\frac{1}{8}$	$\frac{1}{8}$	$\frac{4}{8}$

For example, $\frac{1}{2}$ and $\frac{2}{4}$ are equivalent fractions.

To find equivalent fractions, multiply or divide the numerator and the denominator by a fractional name for 1 such as $\frac{2}{2}$, $\frac{3}{3}$, or $\frac{4}{4}$. Remember, multiplying a number by 1 does not change its value. You are changing the terms, but the value remains the same.

Multiply to get higher terms:

$$\frac{1}{2} = \frac{?}{8} \qquad \boxed{\text{Multiply } \frac{1}{2} \text{ by } \frac{4}{4}, \text{ or } 1.} \qquad \frac{1}{2} = \frac{1 \times 4}{2 \times 4} = \frac{4}{8}$$

Divide to get lower terms:

$$\frac{2}{8} = \frac{?}{4} \qquad \boxed{\text{Divide } \frac{2}{8} \text{ by } \frac{2}{2}, \text{ or } 1.} \qquad \frac{2}{8} = \frac{2 \div 2}{8 \div 2} = \frac{1}{4}$$

simplest form or lowest terms

Since $\frac{2}{8} = \frac{1}{4}$, Phil spent the same amount of time formatting disks and making corrections.

Guided Practice

1. **a.** Rename $\frac{6}{8}$ in higher terms.

$$\frac{6}{8} = \frac{?}{24} \qquad\qquad \frac{6}{8} = \frac{6 \times 3}{8 \times 3} = \underline{\qquad}$$

   b. Rename $\frac{6}{8}$ in lower terms.

$$\frac{6}{8} = \frac{?}{?} \qquad\qquad \frac{6}{8} = \frac{6 \div ?}{8 \div ?} = \underline{\qquad}$$

Complete to rename the fraction in higher terms or lower terms.

2. $\frac{3}{4} = \frac{?}{12}$ _____ 3. $\frac{18}{24} = \frac{?}{8}$ _____ 4. $\frac{4}{5} = \frac{?}{35}$ _____

5. $\frac{8}{20} = \frac{?}{10}$ _____ 6. $\frac{18}{27} = \frac{2}{?}$ _____ 7. $\frac{5}{8} = \frac{15}{?}$ _____

Rename each fraction in simplest form or lowest terms.

8. $\frac{7}{14}$ _____ 9. $\frac{9}{27}$ _____ 10. $\frac{8}{36}$ _____

11. $\frac{20}{32}$ _____ 12. $\frac{16}{80}$ _____ 13. $\frac{21}{98}$ _____

Application

Write a fraction for each and reduce to lowest terms.

14. **a.** What part of an hour is 15 minutes? _____

 b. What part of two weeks is 4 days? _____

 c. What part of 4 days is 8 hours? _____

15. A cubic meter of concrete mix contains 420 kg of cement, 150 kg of stone, and 120 kg of sand. What is the total weight of the mix? What fractional part in lowest terms is cement? stone? sand?

_____

_____

16. I am equivalent to $\frac{2}{3}$. My numerator is 8 less than my denominator. What fraction am I? _____

17. I am equivalent to $\frac{36}{60}$. The sum of my numerator and my denominator is 24. What fraction am I? _____

LEAST COMMON DENOMINATOR

Reminder

$\frac{1}{5}$ ← numerator
← denominator

Vocabulary

least common denominator (LCD): the least common multiple of the denominators of two or more fractions

Reminder

Equivalent fractions are different names for the same part of a whole.

Problem 1: What is the least common denominator of the fractions $\frac{1}{4}$ and $\frac{1}{5}$?

The **least common denominator (LCD)** of two or more fractions is the least common multiple (LCM) of their denominators.

To find the LCD of $\frac{1}{4}$ and $\frac{1}{5}$:

• Find the common multiples of the denominators.

Multiples of 4: 4, 8, 12, 16, (20) 24, . . .

Multiples of 5: 5, 10, 15, (20) 25, 30, . . .

• Find the LCM of the denominators. This is the LCD of the fractions.

The LCD of $\frac{1}{4}$ and $\frac{1}{5}$ is 20.

Problem 2: Write equivalent fractions for $\frac{1}{5}$ and $\frac{2}{3}$ with their least common denominator (LCD).

First find the least common multiple of their denominators: 15

Now multiply to find equivalent fractions with the denominator of 15:

$$\frac{1}{5} = \frac{1 \times 3}{5 \times 3} = \frac{3}{15} \qquad \frac{2}{3} = \frac{2 \times 5}{3 \times 5} = \frac{10}{15}$$

$$LCD = 15$$

Guided Practice

1. Write equivalent fractions for $\frac{5}{6}$ and $\frac{4}{5}$ with their LCD.

   a. Multiples of 6 = 6, 12, _____

   Multiples of 5 = _____

   LCM = _____   LCD = _____

   b. $\frac{5}{6} = \frac{5 \times ?}{6 \times ?} = $ _____

   $\frac{4}{5} = \frac{? \times ?}{? \times ?} = $ _____

Find the least common denominator (LCD) for each pair or group of fractions.

2. $\frac{1}{3}$ and $\frac{1}{2}$ _____

3. $\frac{3}{4}$ and $\frac{2}{5}$ _____

4. $\frac{5}{7}$ and $\frac{7}{9}$ _____

5. $\frac{1}{4}$ and $\frac{3}{8}$ _____

6. $\frac{2}{7}$ and $\frac{5}{21}$ _____

7. $\frac{5}{6}$ and $\frac{3}{10}$ _____

8. $\frac{10}{11}$ and $\frac{9}{44}$ _____

9. $\frac{3}{4}, \frac{5}{6}, \frac{1}{8}$ _____

10. $\frac{7}{12}, \frac{2}{9}, \frac{5}{18}$ _____

Rename each fraction of the pair as an equivalent fraction with the least common denominator (LCD).

11. $\frac{1}{4}$

$\frac{7}{16}$

_____

12. $\frac{3}{5}$

$\frac{5}{8}$

_____

13. $\frac{1}{6}$

$\frac{3}{4}$

_____

14. $\frac{10}{25}$

$\frac{7}{10}$

_____

15. $\frac{3}{4}$

$\frac{9}{10}$

_____

16. $\frac{7}{36}$

$\frac{1}{12}$

_____

17. $\frac{3}{10}$

$\frac{3}{20}$

_____

18. $\frac{1}{9}$

$\frac{1}{11}$

_____

19. $\frac{9}{15}$

$\frac{7}{12}$

_____

Application

20. Ms. Hernandez worked 36 hours on an experiment. She spent 15 hours doing research and 12 hours analyzing data. The rest of the time she spent writing her report. Use fractions to show what part of her time was spent on each task. What did she spend about a quarter of her time doing?

_____

_____

FRACTIONS AND DECIMALS

decimal: a fractional amount written after the decimal point. The decimal point separates the whole number part and the fractional part of a number.

The table shows the part(s) of a dollar that each of four students had left in their pockets as they returned from the amusement park. How can you write these as decimal money amounts?

Name	Dollar(s)
Gaynelle	$\frac{9}{10}$
Chan	$2\frac{7}{100}$
Roy	$\frac{3}{5}$
Adair	$3\frac{1}{4}$

Fractions and mixed numbers with denominators of 10, 100, 1,000, and so on can be renamed as decimals. The word names are the same.

$$\frac{9}{10} = 0.9$$

nine tenths

$$2\frac{7}{100} = 2.07$$

two and seven hundredths

Gaynelle has 0.9 dollar or $0.90. Chan has $2.07.

Reminder

0.2 = 2 tenths
0.02 = 2 hundredths
0.002 = 2 thousandths

To rename fractions like $\frac{3}{5}$ and mixed numbers like $3\frac{1}{4}$ as decimals, divide the numerator by the denominator.

$\frac{3}{5}$ as a ratio means 3 divided by 5.

$$3 \div 5 = 0.6 \qquad \text{Roy has } \$0.60.$$

$$3\frac{1}{4} = 3 + \frac{1}{4} \qquad \text{Divide 1 by 4.}$$

$$1 \div 4 = 0.25 \qquad \text{Adair has } \$3.25.$$

Vocabulary

ratio: a comparison of two numbers by division

Reminder

5 or more, round up; less than 5, round down.

Round decimal answers as needed:

$$\frac{5}{16} = 0.3125 \longrightarrow about\ 0.3$$

nearest tenth

$$\frac{7}{9} = 0.777... \longrightarrow about\ 0.78$$

nearest hundredth

1. Rename $\frac{5}{8}$ and $4\frac{2}{3}$ as decimals.

a. $5 \div 8 =$ _____ $\frac{5}{8} =$ _____

b. _____ $\div$ _____ $=$ _____

$4\frac{2}{3} =$ _____ to the nearest tenth

Exercises

Circle the letter of the equivalent decimal.

2. $\frac{45}{100}$ **a.** 4.500   **b.** 0.45   **c.** 0.045   **d.** 4.05

3. $\frac{7}{1000}$ **a.** 0.07   **b.** 0.070   **c.** 0.700   **d.** 0.007

4. $9\frac{22}{100}$ **a.** 9.22   **b.** 9.022   **c.** 0.922   **d.** 92.20

Rename each fraction or mixed number as a decimal.

5. $\frac{3}{4}$ _____   **6.** $\frac{7}{8}$ _____   **7.** $7\frac{4}{5}$ _____

8. $\frac{6}{15}$ _____   **9.** $\frac{11}{16}$ _____   **10.** $21\frac{8}{32}$ _____

Rename each fraction or mixed number as a decimal. If necessary, round the answer to the nearest hundredth.

11. $\frac{5}{6}$ _____   **12.** $\frac{9}{11}$ _____   **13.** $18\frac{2}{3}$ _____

14. $\frac{17}{25}$ _____   **15.** $7\frac{1}{9}$ _____   **16.** $12\frac{1}{12}$ _____

Application

Use this mental math shortcut to rename fractions as decimals.

- When a fraction has a denominator that is a factor of 10 or 100, rename the fraction to tenths or hundredths.

- Then write the equivalent decimal.

17. $\frac{12}{25} = \frac{12 \times 4}{25 \times 4} = \frac{48}{100} =$ _____   **18.** $\frac{4}{5}$ _____

19. $\frac{3}{20}$ _____   **20.** $\frac{17}{50}$ _____

21. $80\frac{22}{25}$ _____   **22.** $11\frac{24}{40}$ _____

DECIMALS AND FRACTIONS

Some interesting facts:

- A tortoise may travel 0.6 mile in one hour.
- An elephant can run at a speed of 24.5 miles per hour.
- The height of a zebra may be 1.7 yards.
- The height of a flower may be 2.75 feet.

How can you express these numbers using whole numbers and fractions?

Any decimal may be written as a fraction with a denominator of 10, 100, 1000, and so on.

To rename a decimal as a fraction:

- Read the given decimal. $0.6 \longrightarrow$ six tenths

- Determine the denominator of the fraction.

  The denominator is 10.

- Write an equivalent fraction.   six-tenths $= \dfrac{6}{10}$

Reminder

To simplify a fraction, divide by the greatest common factor of the numerator and denominator.

- Rename the fraction in simplest form or lowest terms.

$$\frac{6}{10} = \frac{6 \div 2}{10 \div 2} = \frac{3}{5}$$

$24.5 \longrightarrow$ twenty-four and five tenths

$$\frac{5}{10} = \frac{5 \div 5}{10 \div 5} = \frac{1}{2} \quad \text{Mixed number: } 24\frac{1}{2}$$

$1.7 \longrightarrow$ one and seven tenths

Mixed number: $1\frac{7}{10}$

$2.75 \longrightarrow$ two and seventy-five hundredths

$$\frac{75}{100} = \frac{75 \div 25}{100 \div 25} = \frac{3}{4} \quad \text{Mixed number: } 2\frac{3}{4}$$

Guided Practice

1. Complete. $0.025 = \dfrac{25}{?} = \dfrac{25 \div ?}{?} = $ _____

2. Complete. $9.08 = 9\dfrac{?}{?} = 9\dfrac{? \div ?}{? \div ?} = 9$ _____

Complete.

3. $0.31 = \dfrac{?}{100}$ _____

4. $7.9 = 7\dfrac{?}{10}$ _____

5. $0.8 = \dfrac{8}{?} = \dfrac{?}{5}$ _____

6. $0.057 = \dfrac{57}{?}$ _____

7. $1.23 = 1\dfrac{23}{?}$ _____

8. $8.55 = 8\dfrac{55}{?} = 8\dfrac{?}{?}$ _____

Rename each decimal as a fraction in simplest form.

9. 0.75 _____

10. 0.30 _____

11. 0.2 _____

12. 0.36 _____

13. 0.008 _____

14. 0.675 _____

Rename each decimal as a mixed number in simplest form.

15. 5.4 _____

16. 12.25 _____

17. 9.77 _____

18. 1.44 _____

19. 56.050 _____

20. 4.008 _____

Application

The graph shows the decimal part of the work force that is women.

Estimate the decimal represented by each bar on the graph. Then write each as a fraction in simplest form.

21. Year 2000

_____

22. Year 1990

_____

23. Year 1980

_____

24. Year 1970

_____

25. Year 1960 _____

26. Year 1950 _____

Women in the Work Force: 1950–2000

*Projected

COMPARING FRACTIONS

Anju, Holen, and Cliff each played the same number of games in a tournament. Anju won $\frac{7}{10}$ of her games. Holen won $\frac{2}{3}$ of her games and Cliff won $\frac{4}{5}$ of his games. Arrange the players in order of the number of games won.

To compare fractions with like (or same) denominators, compare their numerators.

$$5 < 7 \text{ so, } \frac{5}{8} < \frac{7}{8}$$

$\frac{5}{8} < \frac{7}{8}$ means $\frac{5}{8}$ is less than $\frac{7}{8}$

Reminder

$\frac{7}{10} = \frac{7 \times 3}{10 \times 3} = \frac{21}{30}.$

$\frac{7}{10}$ and $\frac{21}{30}$ are equivalent fractions.

To compare fractions with unlike denominators, rename them so that the denominators are the same.

- Find the LCD.

$$\frac{7}{10} \quad \frac{2}{3} \quad \frac{4}{5}$$

LCD is 30.

- Rename each fraction using the LCD.

$$\frac{7}{10} = \frac{21}{30} \quad \frac{2}{3} = \frac{20}{30}$$

$$\frac{4}{5} = \frac{24}{30}$$

- Compare the numerators and write the fractions in order.

$$\frac{20}{30} < \frac{21}{30} < \frac{24}{30}$$

From least to greatest: $\frac{2}{3}, \frac{7}{10}, \frac{4}{5}$ Holen, Anju, Cliff

From greatest to least: $\frac{4}{5}, \frac{7}{10}, \frac{2}{3}$ Cliff, Anju, Holen

You can also compare fractions by renaming them as decimals. To rename a fraction as a decimal, divide the numerator by the denominator. Is $\frac{13}{20} > \frac{11}{17}$?

$$\frac{13}{20} \longrightarrow 13 \div 20 = 0.65$$

$$\frac{11}{17} \longrightarrow 11 \div 17 = 0.64706 \text{ (round down to } 0.647)$$

$$0.65 > 0.647. \text{ So, } \frac{13}{20} > \frac{11}{17}.$$

Guided Practice

Compare. Write < or >.

1. $\dfrac{45}{50}$ —————— $\dfrac{21}{50}$

2. $\dfrac{5}{6}$ —————— $\dfrac{4}{5}$

$\downarrow$ $\qquad\qquad\qquad$ $\downarrow$

$\dfrac{?}{30}$ —————— $\dfrac{?}{30}$

Exercises

Compare. Write < , = , or >.

3. $\dfrac{4}{9}$ —————— $\dfrac{7}{9}$ 4. $\dfrac{7}{10}$ —————— $\dfrac{3}{5}$ 5. $\dfrac{3}{5}$ —————— $\dfrac{5}{8}$

6. $\dfrac{11}{11}$ —————— $\dfrac{8}{8}$ 7. $\dfrac{15}{13}$ —————— $\dfrac{9}{13}$ 8. $\dfrac{5}{7}$ —————— $\dfrac{12}{49}$

Write in order from least to greatest.

9. $\dfrac{1}{2}, \dfrac{2}{3}, \dfrac{1}{6}$ 10. $\dfrac{7}{12}, \dfrac{3}{4}, \dfrac{1}{2}$ 11. $\dfrac{11}{18}, \dfrac{7}{9}, \dfrac{2}{3}$

——————————   ——————————   ——————————

Write in order from greatest to least. Then place the fractions on the number line.

12. $\dfrac{7}{9}, \dfrac{1}{3}, \dfrac{4}{6}$ 13. $\dfrac{11}{16}, \dfrac{3}{4}, \dfrac{5}{8}$

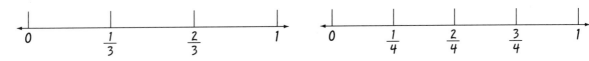

Application

14. Which method of comparing fractions do you prefer? Why?

———————————————————————

———————————————————————

Compare by looking for fractions close to 0, $\dfrac{1}{2}$, or 1.
Write < or >.

15. $\dfrac{6}{13}$ —————— $\dfrac{4}{5}$ 16. $\dfrac{3}{25}$ —————— $\dfrac{27}{50}$ 17. $\dfrac{15}{31}$ —————— $\dfrac{17}{16}$

IMPROPER FRACTIONS AND MIXED NUMBERS

Vocabulary

improper fraction: a fraction with a numerator equal to or greater than its denominator

Reminder

A fraction has a value greater than 1 if the numerator exceeds the denominator. For example, $\frac{5}{6} < 1$; $\frac{6}{6} = 1$; $\frac{7}{6} > 1$.

Reminder

$\frac{18}{6} = 3$, $\frac{24}{6} = 4$, and so on.

Chelsea buys 6 identical bags of rice that weigh a total of 22 pounds. What is the weight of each bag of rice?

The weight of one bag would be $\frac{22}{6}$ pounds. Now rename $\frac{22}{6}$ as a mixed number.

$\boxed{\text{improper fraction}}$

To rename an improper fraction as a whole number or a mixed number:

- Divide the numerator by the denominator.

$$\frac{22}{6} \longrightarrow 6\overline{)22}^{\,3\ R4}$$

- If there is a remainder, write it over the denominator and express in simplest form.

$$\frac{22}{6} = 3\frac{4}{6} = 3\frac{2}{3}$$

So each bag of rice weighs $\frac{22}{6}$ pounds or $3\frac{2}{3}$ pounds.

$\boxed{\text{mixed number}}$

To rename a mixed number as an improper fraction:

- Multiply the whole number by the denominator.

$$2\frac{1}{4} = \frac{(4 \times 2) + 1}{4} = \frac{9}{4}$$

$\boxed{\text{improper fraction}}$

- Add the product to the numerator.

- Write the sum over the denominator.

Guided Practice

1. Express $\frac{11}{3}$ as a mixed number and $6\frac{3}{8}$ as an improper fraction. Locate the numbers on the number line.

a. $\frac{11}{3}$ $3\overline{)11}$ = _____

b. $6\frac{3}{8} = \frac{(8 \times 6) + 3}{8} =$ _____

c.

Express each improper fraction as a whole number or a mixed number in simplest form.

2. $\dfrac{8}{5}$ _____

3. $\dfrac{14}{8}$ _____

4. $\dfrac{12}{12}$ _____

5. $\dfrac{45}{15}$ _____

6. $\dfrac{88}{6}$ _____

7. $\dfrac{92}{10}$ _____

Express each mixed number as an improper fraction.

8. $5\dfrac{3}{4}$ _____

9. $1\dfrac{1}{9}$ _____

10. $5\dfrac{2}{7}$ _____

11. $13\dfrac{1}{3}$ _____

12. $10\dfrac{3}{10}$ _____

13. $2\dfrac{11}{12}$ _____

Locate these numbers on the number line.

14. $\dfrac{42}{10}, \dfrac{8}{5}$

15. $5\dfrac{3}{4}, 8$

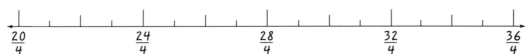

Application

Solve the following.

16. A length of rope is 78 inches long. If it is cut into 8 equal pieces, how long is each piece?

_____

17. Five and three-fourths hours is to be divided into quarter-hour time slots. How many quarter hours can be made?

_____

18. A 2-hour televised telethon is divided into 35-minute segments. Exactly how many segments will there be? How many minutes are left over?

_____

DECIMALS AND MIXED NUMBERS

In a standing long jump contest, three track team members reached these distances: Chico, $6\frac{5}{6}$ ft; Derrick, $6\frac{5}{12}$ ft; and Chi Chen, $6\frac{3}{4}$ ft. Who had the longest jump?

To find who had the longest jump, compare and order the mixed numbers.

You can order the mixed numbers by renaming them so their fractions have the same denominator.

$$6\frac{5}{6} = 6\frac{5 \times 2}{6 \times 2} = 6\frac{10}{12}; \quad 6\frac{3}{4} = 6\frac{3 \times 3}{4 \times 3} = 6\frac{9}{12}$$

From greatest to least: $6\frac{5}{6}, 6\frac{3}{4}, 6\frac{5}{12}$

You can also order mixed numbers by renaming their fraction parts as decimals.

To use decimals to order mixed numbers:

- Rename the fraction part as a decimal.

  $6\frac{5}{6}$ $5 \div 6 = 0.8333333$ $6.83333...$

  $6\frac{5}{12}$ $5 \div 12 = 0.4166666$ $6.41666...$

  $6\frac{3}{4}$ $3 \div 4 = 0.75$ 6.75

- Compare the decimals and write the mixed numbers in order.

  $$6.83333... > 6.75 > 6.416666...$$
  $$\downarrow \qquad\qquad \downarrow \qquad\qquad \downarrow$$
  $$6\frac{5}{6} \quad > \quad 6\frac{3}{4} \quad > \quad 6\frac{5}{12}$$

From greatest to least: $6\frac{5}{6}, 6\frac{3}{4}, 6\frac{5}{12}$

So Chico's jump of $6\frac{5}{6}$ ft was the longest.

Reminder

Reminder

Order numbers by arranging them from greatest to least or least to greatest.

Reminder

Any decimal can be written as a fraction with a denominator of 10, 100, 1000, etc.
$6.75 = 6\frac{75}{100} = 6\frac{3}{4}$

Reminder

To compare decimals, begin comparing in the greatest place:

$0.\textcircled{8}3 > 0.\textcircled{7}5$

$0.\textcircled{7}5 > 0.\textcircled{4}1$

Guided Practice

1. Order $8\frac{1}{3}, 8\frac{5}{6}$, and $8\frac{1}{2}$ from least to greatest by writing equivalent fractions.

   $$8\frac{1}{3} = 8\frac{1 \times 2}{3 \times 2} = 8\frac{?}{6}; \quad 8\frac{1}{2} = 8\frac{1 \times 3}{2 \times 3} = 8\frac{?}{6}$$

   Order: $8\frac{1}{3}$, _____, _____

Reminder

To compare 9.8 and 9.875, rewrite 9.8 as 9.800.
9.800 < 9.875
Adding zeros to the right of the last decimal digit does not change the value of the decimal.

2. Order $9\frac{4}{5}$, $9\frac{7}{8}$, $9\frac{3}{4}$ from greatest to least by renaming them as equivalent decimals.

$4 \div 5 = 0.8$ $9\frac{4}{5} = 9.8$

a. $7 \div 8 = 0.875$ $9\frac{7}{8} = $ _____

b. $3 \div 4 = 0.75$ $9\frac{3}{4} = $ _____

c. Order: _____ , _____ , _____

Exercises

Write in order from greatest to least. Use whichever method you prefer.

3. $1\frac{7}{9}$, $1\frac{5}{6}$, $1\frac{2}{3}$ Order: _____ , _____ , _____

4. $3\frac{7}{18}$, $3\frac{7}{9}$, $3\frac{2}{3}$ Order: _____ , _____ , _____

5. $9\frac{2}{3}$, $9\frac{3}{4}$, $8\frac{4}{5}$ Order: _____ , _____ , _____

6. $7\frac{1}{8}$, $7\frac{1}{10}$, $7\frac{1}{9}$ Order: _____ , _____ , _____

7. $1\frac{2}{11}$, $\frac{14}{11}$, $1\frac{4}{11}$ Order: _____ , _____ , _____

8. $\frac{80}{40}$, $2\frac{23}{40}$, $\frac{13}{5}$ Order: _____ , _____ , _____

Application

Solve the following.

9. José bought three pumpkins at the farmer's market. Their weights were $4\frac{3}{8}$ lb, $4\frac{1}{4}$ lb, and $4\frac{1}{16}$ lb. Which pumpkin was the heaviest? Explain.

_____

_____

10. A friend of yours has missed this lesson. Write an explanation of the two ways you can use to order mixed numbers. Use the numbers in Exercise 9.

_____

_____

_____

ESTIMATING WITH FRACTIONS

Kyela promised to donate at least 12 pounds of food to a homeless shelter. In one bag she has $7\frac{5}{8}$ lb of food. In another bag she has $4\frac{3}{4}$ lb of food. Does Kyela have enough food? You can find out by estimating.

First, round the fraction parts of the mixed numbers.

Estimate if a fraction is close to 0, to $\frac{1}{2}$, or to 1 by comparing its numerator to its denominator.

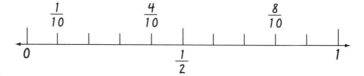

Close to 0: Numerator much less than denominator $\frac{1}{10}$

Close to $\frac{1}{2}$: The numerator doubled about equals the denominator $\frac{4}{10}$

Close to 1: Numerator about equals the denominator $\frac{8}{10}$

So, $\frac{5}{8} \approx \frac{4}{8}$ or $\frac{1}{2}$ $\frac{3}{4} \approx \frac{4}{4}$ or 1

Then add the whole numbers and the rounded fractions.

$$7 + 4 + \frac{1}{2} + 1 = 12\frac{1}{2}$$

Kyela has about $12\frac{1}{2}$ lb of food, enough to keep her promise.

You can also estimate when subtracting fractions and mixed numbers. For example,

$$\frac{3}{5} - \frac{1}{9}$$
$$\downarrow \quad \downarrow$$
$$\frac{1}{2} - 0 = \frac{1}{2}$$

difference: about $\frac{1}{2}$

$$15\frac{1}{8} - 6\frac{4}{7}$$
$$\downarrow \quad \downarrow$$
$$15 - 6\frac{1}{2} = 8\frac{1}{2}$$

difference: about $8\frac{1}{2}$

Guided Practice

Decide if the fractions are closer to 0, $\frac{1}{2}$, or 1. Then estimate the sum or difference.

1. $\frac{8}{10} + \frac{2}{10}$ Estimate:

   $\downarrow$ $\downarrow$ _____

   $1 + 0$

2. $\frac{6}{10}$ $-$ $\frac{1}{10}$ Estimate:

   $\downarrow$ $\downarrow$ _____

   _____ $-$ _____

Estimate the sum or difference.

3. $7\frac{1}{6} \longrightarrow$ 7

   $+ 5\frac{8}{9} \longrightarrow$ $+ 6$ Estimated

   _____   ?   sum: _____

4. $22\frac{3}{4} \longrightarrow$ 23   Estimated

   $- 9\frac{1}{6} \longrightarrow$ $- ?$ difference:

   _____   ?   _____

Exercises

Decide if the fractions are close to 0, $\frac{1}{2}$, or 1. Then estimate the sum or difference.

5. $\frac{2}{15} + \frac{7}{12}$ _____

6. $\frac{8}{9} - \frac{1}{11}$ _____

7. $\frac{1}{8} + \frac{9}{10}$ _____

8. $\frac{10}{12} + \frac{5}{6}$ _____

9. $\frac{7}{15} - \frac{1}{8}$ _____

10. $\frac{17}{19} - \frac{4}{10}$ _____

11. $\frac{7}{8} + \frac{1}{5} + \frac{7}{12}$ _____

12. $\frac{1}{9} + \frac{2}{11} + \frac{1}{10}$ _____

Estimate the sum or difference.

13. $8\frac{3}{4} + 5\frac{4}{5}$ _____

14. $10\frac{1}{6} - 4\frac{7}{15}$ _____

15. $15\frac{1}{2} - \frac{7}{8}$ _____

16. $20\frac{3}{5} + \frac{14}{15} + 9\frac{5}{9}$ _____

Application

COOPERATIVE
LEARNING

17. Write a mixed number that will give a sum or difference in the given range.

   a. $5\frac{2}{9} +$ _____ is between 7 and 8

   b. $7\frac{8}{15} +$ _____ is between 11 and 12

   c. $8\frac{2}{7} -$ _____ is between 3 and 4

   d. $11\frac{4}{7} -$ _____ is between 7 and 8

ADDING FRACTIONS WITH LIKE DENOMINATORS

For strawberry punch, a recipe asks for $\frac{5}{16}$ gal of ginger ale and $\frac{7}{16}$ gal of strawberry cocktail juice. How much liquid is needed? If the recipe is doubled, how much liquid is needed?

$$\frac{5}{16} \qquad\qquad \frac{7}{16}$$

To add fractions with like (same) denominators:

- Add the numerators. $\qquad \frac{5}{16} + \frac{7}{16} = \frac{5+7}{16}$

- Write the result over the common denominator. $\qquad \frac{12}{16}$

- Express the sum in simplest form. $\qquad \frac{12 \div 4}{16 \div 4} = \frac{3}{4}$

Reminder

A fraction is in simplest form when the numerator and denominator have no common factors other than 1.

So $\frac{3}{4}$ gal of liquid is needed for the punch recipe.

To double the recipe, add $\frac{3}{4}$ and $\frac{3}{4}$.

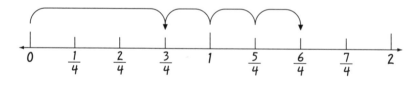

$$\frac{12}{16}, or \frac{3}{4} \qquad\qquad \frac{12}{16}, or \frac{3}{4}$$

$$\frac{3}{4} + \frac{3}{4} = \frac{3+3}{4} = \frac{6}{4} = 1\frac{2}{4}, \ or \ 1\frac{1}{2}$$

So $1\frac{1}{2}$ gallons of liquid are needed to double the recipe.

You can use a number line to add like fractions.

$$
\begin{array}{ccccccccc}
0 & \frac{1}{4} & \frac{2}{4} & \frac{3}{4} & 1 & \frac{5}{4} & \frac{6}{4} & \frac{7}{4} & 2
\end{array}
$$

Guided Practice

1. $\frac{5}{6} + \frac{3}{6} = \frac{8}{6} = $ _____ $= 1\frac{1}{3}$ **2.** $\frac{4}{10} + \frac{7}{10} = \frac{11}{10} = $ _____

Add. Write each answer in simplest form.

3. $\dfrac{4}{16}$
$+\dfrac{2}{16}$

4. $\dfrac{7}{8}$
$+\dfrac{1}{8}$

5. $\dfrac{3}{5}$
$+\dfrac{4}{5}$

6. $\dfrac{5}{24} + \dfrac{7}{24}$ _____

7. $\dfrac{11}{12} + \dfrac{11}{12}$ _____

8. $\dfrac{6}{5} + \dfrac{4}{5}$ _____

9. $\dfrac{2}{15} + \dfrac{8}{15}$ _____

10. $\dfrac{12}{20} + \dfrac{10}{20}$ _____

11. $\dfrac{7}{36} + \dfrac{11}{36}$ _____

12. $\dfrac{16}{25} + \dfrac{12}{25} + \dfrac{22}{25}$ _____

13. $\dfrac{7}{12} + \dfrac{9}{12} + \dfrac{11}{12}$ _____

Find each sum. Use the number lines to help you.

14. $\dfrac{3}{8} + \dfrac{3}{8}$ _____

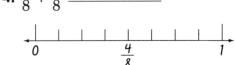

15. $\dfrac{5}{10} + \dfrac{7}{10} + \dfrac{8}{10}$ _____

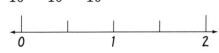

Application

 The fraction strips below show sixteenths. You may want to make your own fraction strips to help you answer.

16. Which fraction strip shows $\dfrac{5}{16}$?

_____

17. Which shows a fraction a little more than $\dfrac{1}{2}$?

_____

18. Name a fraction represented by strip B and E added together.

_____

19. Name a fraction represented by strips C and E added together.

_____

ADDING FRACTIONS WITH UNLIKE DENOMINATORS

Mr. Yoshira added $\frac{1}{2}$ teaspoon of pepper, $\frac{1}{3}$ teaspoon of salt, and $\frac{3}{4}$ teaspoon of curry powder to the stew. How many teaspoons of seasoning were added?

To find the amount of seasoning, find the sum of $\frac{1}{2}$ tsp $+ \frac{1}{3}$ tsp $+ \frac{3}{4}$ tsp. To add fractions with unlike (different) denominators:

* Find the least common denominator (LCD) of the fractions.

  Multiples of 2: 2, 4, 6, 8, 10, (12,) 14, . . .

  Multiples of 3: 3, 6, 9, (12,) 15, 18, 21, . . .

  Multiples of 4: 4, 8, (12,) 16, 20, 24, 28, . . .

  The LCD is 12.

* Rename each fraction as an equivalent fraction with the LCD as the denominator. Then add.

$$\frac{1}{2} = \frac{1 \times 6}{2 \times 6} = \frac{6}{12}$$

$$\frac{1}{3} = \frac{1 \times 4}{3 \times 4} = \frac{4}{12}$$

$$\frac{3}{4} = \frac{3 \times 3}{4 \times 3} = \frac{9}{12}$$

$$\frac{19}{12} = 1\frac{7}{12} \longleftarrow \text{Rename in simplest form.}$$

Estimate to check whether your answer is reasonable.

$$\frac{1}{2} + \frac{1}{3} + \frac{3}{4} \longrightarrow \frac{1}{2} + 0 + 1 \text{ or about } 1\frac{1}{2} \text{ tsp}$$

$1\frac{7}{12} \approx 1\frac{1}{2}$. So, the answer $1\frac{7}{12}$ is reasonable. Mr. Yoshira added $1\frac{7}{12}$ tsp of seasoning.

1. Add $\frac{1}{5}$ and $\frac{3}{20}$.

$$\frac{1}{5} = \frac{1 \times ?}{5 \times 4} = \frac{?}{20}$$ ← Rewrite as an equivalent fraction with LCD, when necessary.

$$+\frac{3}{20} \qquad = \frac{3}{20}$$

$$\frac{7}{20}$$ ← Is the answer in simplest form?

Estimate to check: $\frac{1}{5}$ and $\frac{3}{20}$ are both close to 0. The answer, $\frac{7}{20}$, is less than $\frac{1}{2}$, which is reasonable.

Exercises

Add. Estimate to check whether your answer is reasonable.

2. $\quad \frac{1}{3}$
$\quad +\frac{1}{2}$

3. $\quad \frac{2}{9}$
$\quad +\frac{1}{6}$

4. $\quad \frac{5}{12}$
$\quad +\frac{7}{8}$

5. $\frac{3}{5} + \frac{1}{15}$ _____

6. $\frac{5}{24} + \frac{5}{8}$ _____

7. $\frac{12}{16} + \frac{3}{4}$ _____

8. $\frac{3}{10} + \frac{1}{6}$ _____

9. $\frac{4}{7} + \frac{22}{49}$ _____

10. $\frac{3}{5} + \frac{6}{7}$ _____

11. $\frac{3}{20} + \frac{1}{5} + \frac{3}{10}$ _____

12. $\frac{1}{7} + \frac{3}{14} + \frac{4}{21}$ _____

Application

13. On which day did Ralph and Moniesha swim a total of $\frac{1}{2}$ mile? _____

14. Who swam farther on Wednesday? Thursday? _____

15. Who swam the greater total distance Monday through Thursday?

_____

Distance Swum (miles)		
	Ralph	Moniesha
Mon.	$\frac{1}{10}$	$\frac{1}{5}$
Tues.	$\frac{1}{6}$	$\frac{1}{3}$
Wed.	$\frac{1}{2}$	$\frac{3}{8}$
Thurs.	$\frac{3}{8}$	$\frac{1}{4}$

ADDING MIXED NUMBERS

Reminder

To change an improper fraction like $\frac{6}{4}$ to a mixed number, divide: $6 \div 4$ $= 1\frac{2}{4}$, or $1\frac{1}{2}$

Reminder

Express answers to fraction exercises in simplest form or lowest terms.

Some friends are putting up their own basketball hoop. The pole they are using is $11\frac{3}{4}$ ft high. When attached, the top of the backboard will extend an additional $1\frac{3}{4}$ ft. What is the height of the pole and attached backboard?

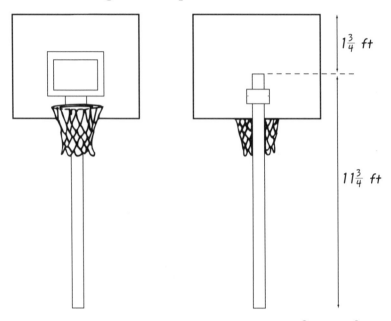

To find the total height, find the sum of $11\frac{3}{4}$ ft + $1\frac{3}{4}$ ft. To add mixed numbers with like (same) denominators:

- Add the fractions

- Add the whole numbers.

$$11\frac{3}{4}$$
$$+\ 1\frac{3}{4}$$
$$12\frac{6}{4} = 12 + 1\frac{1}{2} = 13\frac{1}{2}$$

Estimate to check:

$$11\frac{3}{4} + 1\frac{3}{4}$$
$$12\ +\ 2\ =\ 14\ ft$$

The answer $13\frac{1}{2}$ is reasonable. The height of the pole and attached backboard is $13\frac{1}{2}$ ft.

To add mixed numbers with unlike (different) denominators:

Find the LCD of the fractions. Rename each fraction as an equivalent fraction with the LCD as the denominator. Then add.

$$8\frac{1}{10} = 8\frac{3}{30}$$
$$+\ 12\frac{1}{15} = 12\frac{2}{30}$$
$$20\frac{5}{30},\ or\ 20\frac{1}{6}$$

1.
$$7\frac{5}{8}$$
$$+\,9\frac{1}{8}$$
$$\overline{16\frac{?}{8} = 16\frac{?}{4}}$$

2.
$$5\frac{2}{3} = 5\frac{10}{15}$$
$$+\,3\frac{4}{5} = 3\frac{?}{15}$$
$$\overline{8\frac{?}{15} = 8 + 1\frac{?}{15} = 9\frac{?}{15}}$$

Exercises

Complete. Then give the sum in simplest form.

3.
$$5\frac{1}{4} = 5\frac{?}{8}$$
$$+\,3\frac{3}{8} = 3\frac{3}{8}$$
$$\overline{8\frac{?}{8}}$$ _____

4.
$$2\frac{7}{9}$$
$$+\,11\frac{2}{9}$$
$$\overline{13\frac{?}{?}}$$ _____

5.
$$10\frac{1}{6} = 10\frac{?}{?}$$
$$+\,11\frac{1}{4} = 11\frac{3}{?}$$
$$\overline{21\frac{?}{?}}$$ _____

Add. Estimate to check whether your answer is reasonable.

6.
$$4\frac{3}{4}$$
$$+\,6\frac{1}{4}$$

7.
$$2\frac{1}{5}$$
$$+\,3\frac{1}{3}$$

8.
$$1\frac{1}{2}$$
$$+\,1\frac{1}{2}$$

9.
$$7\frac{5}{18}$$
$$+\,6\frac{1}{6}$$

10.
$$10\frac{2}{7}$$
$$+\,9\frac{5}{6}$$

11.
$$12\frac{7}{10}$$
$$+\,22\frac{7}{30}$$

12. $1\frac{9}{20} + 1\frac{3}{4} =$ _____

13. $18 + 18\frac{7}{12} =$ _____

14. $3\frac{9}{10} + \frac{9}{10} =$ _____

Application

15. Michael has two rectangular picture frames. One is $6\frac{1}{2}$ in. by $8\frac{3}{4}$ in. The other is $9\frac{1}{3}$ in. by 6 in. Which frame has a greater perimeter, or distance around? Explain.

_____

_____

Add. Look for sums of 1 to help you.

16. $8\frac{3}{4} + 5\frac{1}{5} + 11\frac{1}{4}$ _____

17. $2\frac{1}{7} + 3\frac{12}{14} + 5\frac{1}{3}$ _____

SUBTRACTING FRACTIONS WITH LIKE DENOMINATORS

Reminder

Use estimation to check differences as well as sums.

Nanci's sports car has $\frac{5}{8}$ tank of fuel. She makes a road trip that uses $\frac{3}{8}$ tank of fuel. How much fuel will she have left at the end of her trip?

FUEL

To find the remaining fuel, find the difference of $\frac{5}{8}$ tank $-\frac{3}{8}$ tank. To subtract fractions with like (same) denominators:

- Subtract the numerators.

$$\frac{5}{8} - \frac{3}{8} = \frac{5-3}{8}$$

- Write the result over the common denominator.

$$\frac{2}{8}$$

- Express the sum in simplest form.

$$\frac{2}{8} = \frac{2 \div 2}{8 \div 2} = \frac{1}{4}$$

Nanci will have $\frac{1}{4}$ tank of fuel at the end of her trip.

You can use a number line to subtract fractions with like denominators.

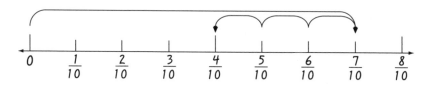

$$\frac{7}{10} - \frac{3}{10} = \frac{7-3}{10} = \frac{4}{10} \text{ or } \frac{2}{5} \text{ in simplest form.}$$

Guided Practice

1. $\frac{14}{16}$ → *about 1*

   $-\frac{7}{16}$ → *about $\frac{1}{2}$*

   $\frac{?}{16}$ → *about* _____

   The difference _____ is about _____. The answer _____ is reasonable.

Exercises

Subtract. Write each answer in simplest form.

2. $\dfrac{4}{5}$
$-\dfrac{1}{5}$

3. $\dfrac{11}{15}$
$-\dfrac{2}{15}$

4. $\dfrac{8}{9}$
$-\dfrac{8}{9}$

5. $\dfrac{10}{12}$
$-\dfrac{8}{12}$

6. $\dfrac{18}{20}$
$-\dfrac{4}{20}$

7. $\dfrac{19}{12}$
$-\dfrac{7}{12}$

8. $\dfrac{20}{32}$
$-\dfrac{12}{32}$

9. $\dfrac{30}{21}$
$-\dfrac{16}{21}$

Find the difference. Label points on the number line to help you.

10. $\dfrac{11}{16} - \dfrac{3}{16}$ _____

Compute. Estimate to check your answer.

11. $\dfrac{15}{16} + \dfrac{3}{16} - \dfrac{7}{16}$ _____

12. $\dfrac{11}{12} - \dfrac{1}{12} - \dfrac{1}{12}$ _____

13. $\dfrac{8}{8} - \dfrac{7}{8} + \dfrac{4}{8} + \dfrac{3}{8}$ _____

14. $\dfrac{10}{22} + \dfrac{4}{22} + \dfrac{8}{22} - \dfrac{18}{22}$ _____

Application

Fifty students chose the environmental concern most important to each of them.

Recycling	✝✝✝✝✝ ✝✝✝✝ ✝✝✝✝ ǀ
Clean Water/Air	✝✝✝✝✝ ✝✝✝✝ ✝✝✝✝ ✝✝✝✝
Disappearing Wildlife	✝✝✝✝✝ ✝✝✝✝
Deforestation	ǀǀǀǀ

15. How much greater is the fraction of students who chose clean water and air than the fraction of those who chose recycling? Show your work.

_____

_____

16. Which fraction is greater: the fraction of those who chose clean water and air or the fraction of students who chose the other three concerns? How much greater?

_____

SUBTRACTING FRACTIONS WITH UNLIKE DENOMINATORS

Francisco, a scrap metal dealer, has a stack of $\frac{3}{4}$-yd copper bars. A client orders a supply of $\frac{2}{3}$-yd bars. What length must Francisco cut off each $\frac{3}{4}$-yd bar to fill the order?

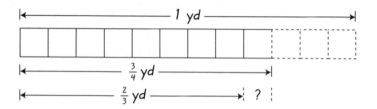

To find the length of each bar that must be cut off, find the difference between $\frac{3}{4}$ yd and $\frac{2}{3}$ yd. To subtract fractions with unlike (different) denominators:

Reminder

The second, or bottom, term of a fraction is the denominator.

- Find the lowest common denominator (LCD) of the fractions.

Multiples of 3:   3, 6, 9, ⑫, 15, 18, ...

Multiples of 4:   4, 8, ⑫, 16, 20, 24, ...

The LCD is 12.

- Rename each fraction as an equivalent fraction with the LCD as the denominator. Then subtract.

$$\frac{3}{4} - \frac{2}{3} \longrightarrow \frac{9}{12} - \frac{8}{12} = \frac{9-8}{12} = \frac{1}{12}$$

So $\frac{1}{12}$ yd of each bar must be cut off to fill the order.

Guided Practice

1. $\frac{4}{5} = \frac{4 \times ?}{5 \times ?} = \frac{?}{15}$ ⟵ Write equivalent fractions with LCD. Subtract.

$-\frac{1}{15} \qquad = \frac{1}{15}$

$\overline{\qquad ? \qquad}$

Reminder

Estimate to check whether your answer is reasonable.

2. $\frac{3}{8} - \frac{1}{10} \longrightarrow \frac{15}{?} - \frac{4}{?} = $ _____

Exercises

Subtract. Write each answer in simplest form.

3.
$$\frac{1}{2}$$
$$-\frac{1}{3}$$

4.
$$\frac{5}{6}$$
$$-\frac{1}{12}$$

5.
$$\frac{2}{5}$$
$$-\frac{2}{15}$$

6.
$$\frac{6}{7}$$
$$-\frac{5}{14}$$

7.
$$\frac{5}{6}$$
$$-\frac{1}{10}$$

8.
$$\frac{4}{12}$$
$$-\frac{12}{36}$$

9. $\frac{5}{8} - \frac{3}{8}$ _____

10. $\frac{6}{7} - \frac{3}{4}$ _____

11. $\frac{6}{8} - \frac{1}{6}$ _____

12. $\frac{7}{6} - \frac{3}{5}$ _____

13. $\frac{10}{9} - \frac{1}{8}$ _____

14. $\frac{5}{12} - \frac{8}{25}$ _____

Add or subtract as indicated.

15. $\frac{1}{2} + \frac{1}{3} - \frac{1}{4}$ _____

16. $\frac{3}{5} - \frac{1}{3} + \frac{3}{4}$ _____

Application

17. Five 6-inch strips are cut from a piece of plywood that is $1\frac{2}{3}$ yd long. How long is the piece of plywood that is left? (Hint: 12 in. = 1 ft = $\frac{1}{3}$ yd)

_____

18. A recent survey shows that the people of Mexico spend $\frac{1}{3}$ of their family income on food, the people of the United States spend $\frac{1}{10}$ of their income on food, and the people of India spend $\frac{11}{20}$.

a. How much greater is the fractional part for India than for the United States? _____

b. Which is greater: the fractional part for Mexico or for India? How much greater? _____

SUBTRACTING MIXED NUMBERS

Problem 1: Jesus and Mayra work weekends making silk flowers to sell at street festivals. To make a rose, they use $12\frac{1}{8}$ in. of silk. To make a lily, they use $16\frac{7}{8}$ in. of silk. How much more silk do they use to make a lily than a rose?

To find how much more, subtract: $16\frac{7}{8}$ in. $- 12\frac{1}{8}$ in. To subtract mixed numbers with like (same) denominators:

Subtract the fractions.

$$16\frac{7}{8}$$

Reminder

Express all answers in simplest form.

Reminder

$16\frac{7}{8}$ rounds up to the nearest whole number; $12\frac{1}{8}$ rounds down.

Subtract the whole numbers.

$$-\,12\frac{1}{8}$$
$$\overline{\qquad\quad 4\frac{6}{8} = 4\frac{3}{4}}$$

Estimate to check: $16\frac{7}{8} - 12\frac{1}{8} \longrightarrow 17 - 12 = 5$

About 5 in. more. The answer is reasonable.

They use $4\frac{3}{4}$ in. more silk to make a lily.

Problem 2: Jesus cuts wire stems for the flowers. From a piece of gold wire $11\frac{1}{2}$ in. long, he cuts a stem $9\frac{1}{16}$ in. long. How much gold wire is left?

To find how much wire is left, subtract: $11\frac{1}{2} - 9\frac{1}{16}$. To subtract mixed numbers with unlike (different) denominators:

- Find the LCD of the fractions. $11\frac{1}{2} = 11\frac{8}{16}$

- Rename each fraction as an equivalent fraction with the LCD as the denominator.

$$-\,9\frac{1}{16} = 9\frac{1}{16}$$
$$\overline{\qquad\quad 2\frac{7}{16}}$$

- Subtract the fractions.

- Subtract the whole numbers.

$2\frac{7}{16}$ in. of wire is left.

Reminder

$9\frac{1}{16}$ rounds to 9.

Estimate to check:

$$11\frac{1}{2} - 9\frac{1}{16} \longrightarrow 11\frac{1}{2} - 9 = 2\frac{1}{2}$$

The difference is about $2\frac{1}{2}$. The answer is reasonable.

1. $7\frac{4}{5}$
     $-3\frac{2}{5}$
     $\overline{4\frac{?}{5}}$

2. $11\frac{2}{3} = 11\frac{?}{6}$ ← Write as equivalent fractions with LCD.
     $-5\frac{1}{6} = 5\frac{1}{6}$ Subtract.
     $\overline{6\frac{?}{6} = 6\frac{?}{2}}$ ← Rewrite in lowest terms.

Exercises

Subtract. Estimate to check whether your answers are reasonable.

3. $8\frac{4}{15}$
     $-1\frac{2}{15}$

4. $10\frac{10}{11}$
     $-5\frac{5}{11}$

5. $9\frac{3}{4}$
     $-2\frac{1}{4}$

6. $9\frac{7}{12}$
     $-3\frac{5}{12}$

7. $18\frac{4}{9}$
     $-4\frac{2}{9}$

8. $6\frac{1}{2}$
     $-5\frac{1}{4}$

9. $11\frac{2}{3}$
     $-3\frac{1}{5}$

10. $19\frac{6}{15}$
      $-11\frac{1}{4}$

11. $6\frac{5}{18}$
      $-2\frac{1}{5}$

12. $12\frac{1}{9} - 6 =$ _____

13. $5\frac{7}{8} - 5\frac{7}{8} =$ _____

Application

14. Mia and Omar are skating along an $8\frac{3}{4}$-mile path. They reach a marker that says the end of the path is $5\frac{1}{2}$ miles away. How far have they skated? Estimate how much farther they would have to go in order to skate 20 miles.

_____

15. Cleon wants to make two kinds of bread for a pot luck dinner. He has $4\frac{1}{2}$ cups of flour. One recipe calls for $2\frac{3}{4}$ cups of flour. If he uses this recipe, will he have enough flour to make another recipe that calls for $1\frac{1}{2}$ cups of flour? How do you know?

_____

SUBTRACTING MIXED NUMBERS WITH RENAMING

Part-time Work Schedule		
Day	Rohan	Elena
Sunday	$6\frac{1}{2}$ h	7 h
Monday	6 h	$5\frac{1}{2}$ h
Tuesday	$2\frac{3}{4}$ h	8 h
Wednesday	8 h	—
Thursday	$3\frac{1}{2}$ h	$6\frac{1}{4}$ h
Friday	$5\frac{1}{2}$ h	5 h

On Thursday, how many more hours did Elena work than Rohan?

To find how many more hours, you need to subtract.

$$6\frac{1}{4} = 6\frac{1}{4}$$
$$- 3\frac{1}{2} = 3\frac{2}{4}$$

The fractional parts of mixed numbers cannot be subtracted
$$\frac{2}{4} > \frac{1}{4}$$

Rename the $6\frac{1}{4}$ with an improper fraction so that you can subtract the fractional parts.

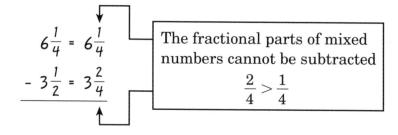

$$6\frac{1}{4} = 5\frac{5}{4}$$
$$- 3\frac{2}{4} = 3\frac{2}{4}$$
$$2\frac{3}{4}$$

$$6\frac{1}{4} = 5 + 1 + \frac{1}{4}$$
$$= 5 + \frac{4}{4} + \frac{1}{4}, \text{ or } 5\frac{5}{4}$$

So Elena worked $2\frac{3}{4}$ hours more on Thursday than Rohan.

Complete.

Reminder

Sometimes you will not know whether fractional parts can be subtracted until you write equivalent fractions with a common denominator.

1. $7\frac{2}{5} = 6 + 1 + \frac{2}{5}$

   $= 6 + \frac{?}{5} + \frac{2}{5} = 6\frac{?}{5}$

2. $12 = 11 + 1$

   $= 11 + \frac{?}{8} = 11\frac{?}{8}$

3. $9\frac{3}{4} = 9\frac{9}{12} = 8\frac{?}{12}$

   $-4\frac{5}{6} = 4\frac{10}{12} = 4\frac{?}{12}$
   _____
   $4\frac{?}{12}$

4. $8 = 7\frac{?}{?}$

   $-4\frac{1}{5} = 4\frac{1}{5}$
   _____
   $3\frac{?}{?}$

Exercises

Subtract. Estimate to check your answer.

5. $5\frac{1}{3}$
   $-2\frac{2}{3}$
   _____

6. $7\frac{1}{8}$
   $-3\frac{3}{8}$
   _____

7. $8\frac{1}{3}$
   $-1\frac{5}{6}$
   _____

8. $10\frac{1}{6}$
   $-4\frac{3}{4}$
   _____

9. 9
   $-1\frac{1}{2}$
   _____

10. 6
   $-3\frac{2}{3}$
   _____

11. $6\frac{1}{5}$
   $-2\frac{3}{4}$
   _____

12. $13\frac{1}{12}$
   $-11\frac{9}{10}$
   _____

13. $15\frac{1}{5}$
   $-6\frac{3}{8}$
   _____

Application

Answer each question. Use the "Part-Time Work Schedule" on the opposite page.

14. How many more hours did Rohan work on Sunday than on Tuesday? On Wednesday than on Tuesday?

_____

_____

MULTIPLYING FRACTIONS

One-third of the eighth graders at Lansing Junior High are bilingual. Three-fourths of these bilingual students speak Spanish. What part of the eighth graders speak Spanish?

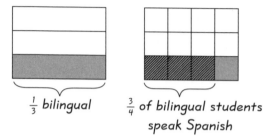

$\frac{1}{3}$ bilingual $\frac{3}{4}$ of bilingual students speak Spanish

Find the product of $\frac{3}{4} \times \frac{1}{3}$

To multiply a fraction by a fraction:

- Multiply the numerators. $\frac{3}{4} \times \frac{1}{3} = \frac{3 \times 1}{4 \times 3}$

- Multiply the denominators.

- Write the product in simplest form. $= \frac{3}{12} = \frac{1}{4}$

So $\frac{1}{4}$ of the eighth graders speak Spanish.

You can estimate to check products. The product of a fraction times a fraction is less than either fraction. The product of a whole number and a fraction is less than the whole number. $\frac{1}{4}$ is less than $\frac{1}{3}$ and $\frac{3}{4}$, so the answer is reasonable.

To help multiply a fraction by a fraction or a fraction by a whole number, you can simplify terms before multiplying.

$$\frac{\overset{1}{\cancel{3}}}{4} \times \frac{1}{\underset{1}{\cancel{3}}} \qquad \begin{array}{l}\text{Divide any numerator or denominator} \\ \text{by their greatest common factor.} \\ \text{Rewrite the terms you divided.}\end{array}$$

$$\frac{\overset{1}{\cancel{7}}}{\underset{2}{\cancel{12}}} \times \frac{\overset{1}{\cancel{6}}}{\underset{1}{\cancel{7}}} = \frac{1 \times 1}{2 \times 1} = \frac{1}{2}$$

Check: $\frac{1}{2}$ is less than $\frac{7}{12}$ and $\frac{6}{7}$. The answer is reasonable.

Reminder

Multiplying a number by a proper fraction will produce a smaller number. For example, 8, or $\frac{8}{1} \times \frac{1}{2} = \frac{8}{2}$ or 4. Eight halves is the same as four wholes.

$$7 \times \frac{5}{21} = \frac{\overset{1}{\cancel{7}}}{1} \times \frac{5}{\underset{3}{\cancel{21}}} = \frac{5}{3} = 1\frac{2}{3}$$

Check: $1\frac{2}{3}$ is less than 7. The answer is reasonable.

Guided Practice

1. $\dfrac{3}{5} \times \dfrac{7}{8} = \dfrac{3 \times ?}{5 \times 8} = \dfrac{?}{?}$

2. $\dfrac{7}{20} \times 10 = \dfrac{7}{\underset{?}{\cancel{20}}} \times \dfrac{\overset{1}{\cancel{10}}}{1} = \dfrac{?}{2} = $ _____

Exercises

Multiply. Simplify before multiplying when possible.

3. $\dfrac{2}{5} \times \dfrac{5}{8}$ _____

4. $\dfrac{5}{12} \times \dfrac{3}{4}$ _____

5. $\dfrac{7}{9} \times \dfrac{2}{3}$ _____

6. $\dfrac{6}{7} \times \dfrac{2}{5}$ _____

7. $\dfrac{1}{3} \times 30$ _____

8. $18 \times \dfrac{5}{6}$ _____

9. $\dfrac{1}{2} \times \$36$ _____

10. $\dfrac{2}{3}$ of $\$45$ _____

11. $\dfrac{1}{4}$ of $\$9.60$ _____

12. $\dfrac{3}{5} \times \dfrac{2}{9} \times 20$ _____

13. $\dfrac{7}{8} \times \dfrac{2}{3} \times \dfrac{1}{14}$ _____

Application

14. Ophir's goal is to collect $95 for the walkathon. She has collected $\dfrac{3}{5}$ of that amount. How much has she collected?

_____

15. Of the students in the eighth grade at Lansing Junior High, $\dfrac{2}{5}$ are Asian Americans. $\dfrac{1}{4}$ of the Asian Americans are of Chinese descent. What fractional part of the students in the eighth grade are of Chinese descent? What fractional part are not?

_____

_____

MULTIPLYING MIXED NUMBERS

A running path around a lake is $2\frac{1}{4}$ miles long. If Abe completes $7\frac{1}{3}$ laps on the path, how many miles will he have run?

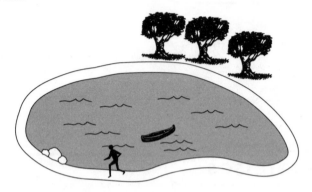

Find the product of: $2\frac{1}{4} \times 7\frac{1}{3}$.

To multiply mixed numbers:

- Rename both mixed numbers as improper fractions.

$$2\frac{1}{4} \times 7\frac{1}{3} = \frac{9}{4} \times \frac{22}{3}$$

Reminder

$$2\frac{1}{4} = \frac{4 \times 2 + 1}{4} = \frac{9}{4}$$

- Simplify terms where possible.

$$= \frac{\overset{3}{\cancel{9}}}{\underset{2}{\cancel{4}}} \times \frac{\overset{11}{\cancel{22}}}{\underset{1}{\cancel{3}}}$$

- Multiply the numerators.
- Multiply the denominators.

$$= \frac{3 \times 11}{2 \times 1}$$

- If the product is an improper fraction, rename it as a whole number or mixed number.

$$= \frac{33}{2} = 16\frac{1}{2}$$

Abe will have run $16\frac{1}{2}$ miles.

Estimate to check: $2\frac{1}{4} \times 7\frac{1}{3} \longrightarrow 2 \times 7$, or about 14. The answer is reasonable.

Reminder

$\frac{1}{3} \longrightarrow 1 \div 3$, or 0.33 to the nearest one-hundredth.

You can also use a calculator to find or check the answer. Rename the mixed numbers as decimals.

$$2.25 \times 7.33 = 16.4925$$

Notice that calculator answers are not exact if you input a rounded value for a fraction.

1. $6\frac{1}{4} \times 2\frac{2}{5} = \frac{?}{4} \times \frac{?}{5}$

$= \frac{\overset{5}{\cancel{25}}}{\underset{1}{\cancel{4}}} \times \frac{\overset{?}{\cancel{12}}}{\underset{1}{\cancel{5}}}$

$= \frac{5 \times ?}{1 \times ?} = $ _____

2. $3 \times 2\frac{1}{9} = \frac{3}{1} \times \frac{?}{?}$

$= \frac{\overset{1}{\cancel{3}}}{1} \times \frac{19}{\underset{3}{\cancel{9}}}$

$= \frac{1 \times ?}{1 \times ?} = $ _____

Exercises

Multiply. Estimate to check the answer.

3. $1\frac{1}{2} \times 1\frac{1}{3}$ _____

4. $1\frac{1}{5} \times 1\frac{1}{6}$ _____

5. $10 \times 3\frac{1}{5}$ _____

6. $4\frac{1}{11} \times 22$ _____

7. $3\frac{3}{4} \times 1\frac{2}{5}$ _____

8. $1\frac{1}{5} \times 3\frac{3}{4}$ _____

9. $4\frac{1}{2} \times \frac{5}{9}$ _____

10. $\frac{3}{4}$ of $2\frac{2}{3}$ _____

11. $\frac{5}{6}$ of $7\frac{1}{2}$ _____

12. $1\frac{1}{2} \times 2\frac{1}{4} \times 1\frac{1}{3}$ _____

13. $3\frac{1}{8} \times 4 \times \frac{3}{10}$ _____

Use your calculator to find the products. Then compare.
Write < or >

14. $7\frac{1}{2} \times 2\frac{1}{4}$ ____?____ $3\frac{1}{4} \times 5\frac{1}{2}$

15. $4\frac{1}{2} \times 6\frac{2}{5}$ ____?____ $11\frac{1}{10} \times 2\frac{1}{5}$

16. $1\frac{2}{3} \times 2\frac{1}{2}$ ____?____ $2\frac{2}{5} \times 1\frac{1}{2}$

17. $4\frac{1}{2} \times 20\frac{1}{8}$ ____?____ $40\frac{1}{2} \times 2\frac{3}{8}$

Application

18. A photograph measures $3\frac{1}{2}$ inches wide. If the photograph were enlarged $2\frac{1}{2}$ times, how wide would it be?

_____

19. A long-playing record makes $33\frac{1}{3}$ revolutions per minute. About how many revolutions does it make in one-half hour?

_____

DIVIDING FRACTIONS

Vocabulary

reciprocal: one of any 2 numbers whose product is 1. For example, 3 (or $\frac{3}{1}$) and $\frac{1}{3}$ are reciprocals. $\frac{7}{8}$ and $\frac{8}{7}$ are reciprocals.

Reminder

The number you are dividing by is the divisor. The number you divide is the dividend. The answer is the quotient.

$$\frac{1}{2} \div \frac{1}{8} = 4$$

dividend   divisor   quotient

Problem 1: Latoya cut a $\frac{1}{2}$-yd bolt of cloth into $\frac{1}{8}$-yd strips. How many strips did she make?

To find the answer, you can model how many $\frac{1}{8}$-strips there are in $\frac{1}{2}$.

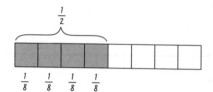

$$\frac{1}{8} + \frac{1}{8} + \frac{1}{8} + \frac{1}{8}, \text{ or}$$

$$4 \times \frac{1}{8} = \frac{1}{2}$$

Latoya made 4 strips of cloth.

One way to divide fractions is to multiply the dividend by the **reciprocal** of the divisor. Two numbers whose product is 1 are reciprocals. For example, $\frac{3}{8}$ and $\frac{8}{3}$ are reciprocal because

$$\frac{3}{8} \times \frac{8}{3} = \frac{3 \times 8}{8 \times 3} = \frac{24}{24} = 1$$

Find the reciprocal of any fraction by exchanging the numerator and denominator. What is the reciprocal of 5?

$$5 = \frac{5}{1} \quad \frac{1}{5}$$

The reciprocal of 5 is $\frac{1}{5}$.

Check: $\frac{5}{1} \times \frac{1}{5} = \frac{5 \cdot 1}{1 \cdot 5} = \frac{5}{5} = 1$

You can also find the answer to Problem 1 using division. To divide a fraction by a fraction:

- Multiply by the reciprocal of the divisor:

$$\frac{1}{2} \div \frac{1}{8} = \frac{1}{2} \times \frac{8}{1}$$

- Simplify whenever possible.

$$= \frac{1}{\overset{}{\underset{1}{2}}} \times \frac{\overset{4}{8}}{1}$$

- Multiply the numerators. Multiply the denominators.

$$= \frac{1 \times 4}{1 \times 1}$$

- If the quotient is an improper fraction, rename it.

$$= \frac{4}{1} = 4$$

Divide fractions and whole numbers in a similar way.

Reminder

When dividing fractions, express whole numbers as improper fractions with denominators of 1.

Problem 2: $4 \div \dfrac{3}{5} = \dfrac{4}{1} \times \dfrac{5}{3}$

$$= \dfrac{4 \times 5}{1 \times 3} = \dfrac{20}{3}, \text{ or } 6\dfrac{2}{3}$$

Problem 3: $\dfrac{3}{5} \div 4 = \dfrac{3}{5} \times \dfrac{1}{4}$

$$= \dfrac{3}{20}$$

Guided Practice

1. $\dfrac{6}{7} \div \dfrac{3}{4} = \dfrac{6}{7} \times \dfrac{?}{?}$ Multiply by the reciprocal.

   $= \dfrac{?}{?}$ Simplify terms if possible.

   $= 1\dfrac{?}{?}$ Write answer as a mixed number.

2. $\dfrac{2}{5} \div 10 = \dfrac{2}{5} \times \dfrac{1}{?} = \dfrac{1}{?}$

Exercises

Divide.

3. $\dfrac{3}{5} \div \dfrac{1}{5}$ _____

4. $\dfrac{4}{9} \div \dfrac{5}{9}$ _____

5. $\dfrac{2}{3} \div \dfrac{3}{4}$ _____

6. $\dfrac{1}{9} \div \dfrac{1}{18}$ _____

7. $12 \div \dfrac{3}{2}$ _____

8. $3 \div \dfrac{2}{7}$ _____

9. $\dfrac{1}{4} \div 6$ _____

10. $\dfrac{11}{8} \div \dfrac{8}{8}$ _____

11. $\dfrac{2}{3} \div 8$ _____

Application

12. How many $\dfrac{1}{8}$-in. leather strips can be cut from a $\dfrac{3}{4}$-in. piece of leather?

_____

13. How many $\dfrac{2}{3}$-cup sugar bowls can be filled from 16 cups of sugar?

_____

DIVIDING MIXED NUMBERS

Reminder

The product of a number and its reciprocal is 1:

$\frac{4}{5} \times \frac{5}{4} = 1$.

The Big Apple Bikers' Club is organizing a $6\frac{1}{4}$-hour trip. A rest stop is planned after every $1\frac{1}{4}$ hours of biking. How many rest stops will there be?

Find the quotient of: $6\frac{1}{4} \div 1\frac{1}{4}$.

To divide mixed numbers:

- Rename both mixed numbers as improper fractions.

$$6\frac{1}{4} \div 1\frac{1}{4} = \frac{25}{4} \div \frac{5}{4}$$

- Multiply by the reciprocal of the divisor.

$$= \frac{25}{4} \times \frac{4}{5}$$

- Simply whenever possible.

$$= \frac{\overset{5}{\cancel{25}}}{\underset{1}{\cancel{4}}} \times \frac{\overset{1}{\cancel{4}}}{\underset{1}{\cancel{5}}}$$

- Multiply the numerators. Multiply the denominators.

$$= \frac{5 \times 1}{1 \times 1}$$

- If the quotient is an improper fraction, rename it as a whole number or mixed number.

$$= \frac{5}{1} = 5$$

The Bikers' Club will make 5 rest stops.

Estimate to check: $6\frac{1}{4} \div 1\frac{1}{4} \longrightarrow 6 \div 1 = 6$

About 6 rest stops. The answer is reasonable.

You can use a calculator to find the answer. Rename the mixed numbers as decimals.

$$6 . 2 5 \; \boxed{\div} \; 1 . 2 5 \; \boxed{=} \; 5 .$$

Reminder

Calculator answers are not exact if you input a rounded value for a fraction.

Divide mixed numbers and whole numbers in a similar way.

$$10\frac{1}{2} \div 9 = \frac{21}{2} \div \frac{9}{1} = \frac{21}{2} \times \frac{1}{9}$$

$$= \frac{\overset{7}{\cancel{21}}}{2} \times \frac{1}{\underset{3}{\cancel{9}}} = \frac{7}{6} = 1\frac{1}{6}$$

1. $2\frac{1}{4} \div 1\frac{1}{4} = \frac{9}{4} \div \frac{5}{4}$

   $= \frac{9}{4} \times \frac{?}{?} = $ _____

2. $3\frac{1}{2} \div 14 = \frac{7}{2} \div \frac{?}{?}$

   $= \frac{7}{2} \times \frac{?}{?} = $ _____

Exercises

Divide. Estimate to check the answer.

3. $3\frac{1}{3} \div 1\frac{2}{3}$

_____

4. $10\frac{1}{2} \div 1\frac{1}{2}$

_____

5. $1\frac{1}{8} \div 3\frac{1}{2}$

_____

6. $4\frac{1}{5} \div 4\frac{2}{3}$

_____

7. $2\frac{2}{5} \div 1\frac{7}{10}$

_____

8. $1\frac{1}{3} \div 6$

_____

9. $12 \div 2\frac{1}{2}$

_____

10. $15 \div 1\frac{1}{4}$

_____

11. $7\frac{1}{5} \div 9$

_____

12. $\frac{5}{6} \div 2\frac{1}{7}$

_____

13. $3\frac{3}{5} \div \frac{9}{10}$

_____

14. $2\frac{3}{4} \div 2\frac{6}{8}$

_____

Compute. Do operations within parentheses first.

15. $\left(3\frac{1}{2} \div 6\right) - \frac{1}{4}$ _____

16. $\left(\frac{5}{16} + \frac{3}{8}\right) \times 8$ _____

17. $\left(7\frac{1}{2} \div 1\frac{1}{4}\right) \times 1\frac{1}{8}$ _____

18. $\left(5\frac{1}{6} - 1\frac{2}{3}\right) \div 4$ _____

Application

19. It takes 48 inches of material to make one shirt. How many shirts can be made from $16\frac{1}{2}$ yards of material?

_____

_____

20. The local movie theater has 14 hours and 40 minutes on the weekend to show a 1 hour and 45 minute movie. How many times can the entire movie be shown? About how much time will be left over? (Hint: change times to mixed numbers first.)

_____

_____

FRACTIONS AND PERCENTS

Vocabulary

percent: the ratio of a number to 100. 29% means 29:100 or 29 parts per 100.

Raoul took a survey of 50 students. He asked whether they approved of the job that each class officer was doing. The "yes" responses for the 4 class officers are shown on the graph. How can Raoul express each officer's results as a percent score, or approval rating?

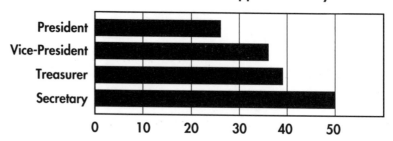

Class Officer Approval Survey

The % sign stands for percent. One definition of **percent** is "a given part compared to one hundred parts."

To express the survey results as percent scores, you must rename the fractions as ratios to 100.

Reminder

$\frac{80}{100}$ is read as "eighty hundredths" or "eighty over one hundred."

President: $\quad \frac{26}{50} = \frac{26 \times 2}{50 \times 2} = \frac{52}{100}$ or 52%

Vice-President: $\quad \frac{36}{50} = \frac{36 \times 2}{50 \times 2} = \frac{72}{100}$ or 72%

Treasurer: $\quad \frac{39}{50} = \frac{39 \times 2}{50 \times 2} = \frac{78}{100}$ or 78%

Secretary: $\quad \frac{50}{50} = \frac{50 \times 2}{50 \times 2} = \frac{100}{100}$ or 100%

100% of something means all of it.

52% of the students surveyed approve of the job that the class president is doing. The vice-president's approval rating is 72%, the treasurer's rating is 78%, and the secretary's rating is 100%.

To rename a fraction as a percent:

Reminder

Equivalent fractions have the same value.

- Express the fraction as an equivalent fraction with a denominator of 100.

$\frac{3}{5} = \frac{3 \times 20}{5 \times 20} = \frac{60}{100}$

- Write this number of hundredths as a percent.

$\frac{60}{100} = 60\%$

You can use a calculator to rename fractions as percents.

$\frac{3}{20}$ 3 ÷ 20 % 0.15   15%

$\frac{7}{16}$ 7 ÷ 16 % 0.4375   43.75%

Guided Practice

Reminder

When expressing ratios as percents, round to the nearest hundredth.

1. $\frac{15}{100} = $ _____?_____ %

2. $\frac{1}{25} = \frac{1 \times ?}{25 \times 4} = \frac{?}{?} = $ _____?_____ %

3. $\frac{5}{8}$ 5 ÷ _____?_____ % = _____?_____ %

Exercises

Rename the fraction as a percent.

4. $\frac{24}{100}$ _____

5. $\frac{8}{100}$ _____

6. $\frac{3}{4}$ _____

7. $\frac{2}{5}$ _____

8. $\frac{7}{25}$ _____

9. $\frac{3}{10}$ _____

10. $\frac{11}{20}$ _____

11. $\frac{1}{25}$ _____

12. $\frac{15}{15}$ _____

 Use a calculator to rename each fraction as a percent.

13. $\frac{7}{8}$ _____

14. $\frac{5}{16}$ _____

15. $\frac{3}{5}$ _____

16. $\frac{22}{32}$ _____

17. $\frac{2}{3}$ _____

18. $\frac{5}{9}$ _____

Application

19. The table shows the results of a department store survey of reasons customers come to the store. Find each percent.

Reason	Normal	%	Advertising Day	%
Advertising	20		72	
Regular Customer	55		24	
Friend Recommends	9		6	
Driving By	6		18	
TOTALS	90		120	

 **PERCENTS AND FRACTIONS**

A recent study was completed to find out what percent of the athletic shoes imported into the United States come from the various countries in Asia. The circle graph shows the results.

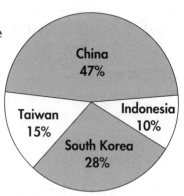

From what country did about one-half of the shoes come? From what country did about one-quarter of the shoes come?

You can use the following "benchmarks" to help you estimate with percents.

Reminder

Percent means "the part of each hundred."

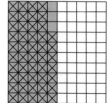 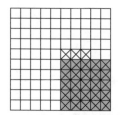 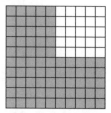

$$50\% = \frac{50}{100} = \left(\frac{1}{2}\right) \qquad 25\% = \frac{25}{100} = \left(\frac{1}{4}\right) \qquad 75\% = \frac{75}{100} = \left(\frac{3}{4}\right)$$

Percents around 50% are about $\frac{1}{2}$.   Percents around 25% are about $\frac{1}{4}$.   Percents around 75% are about $\frac{3}{4}$.

$$47\% = \frac{47}{100} \approx \frac{1}{2} \qquad 28\% = \frac{28}{100} \approx \frac{1}{4}$$

So about $\frac{1}{2}$ of the shoes came from China, and about $\frac{1}{4}$ came from South Korea.

To rename a percent as a fraction:

- Drop the percent sign (%). $15\% \longrightarrow 15$

- Write the number as the numerator and 100 as the denominator. $\frac{15}{100}$

- Express the fraction in simplest form. $\frac{15}{100} = \frac{15 \div 5}{100 \div 5} = \frac{3}{20}$

So, $\frac{3}{20}$ of the athletic shoes imported from Asia come from Taiwan.

Guided Practice

1. Rename 44% as a fraction in simplest form.

    a. Write the number in hundredths. $\qquad 44\% = \frac{?}{100}$

    b. Express the fraction in simplest form. $\qquad \frac{?}{100} = \frac{? \div 4}{100 \div 4} = \frac{?}{25}$

Exercises

Rename the percent as a fraction in simplest form.

2. 60% _____

3. 90% _____

4. 5% _____

5. 35% _____

6. 72% _____

7. 8% _____

8. 13% _____

9. 27% _____

10. 88% _____

11. In questions 2–10, which percent is about $\frac{3}{4}$? About $\frac{1}{4}$?

_____

Rewrite each statement with a fraction in simplest form.

12. More than 92% of the basketball players in the N.B.A. are over 6 ft.

_____

13. About 64% of the ticket holders went to see the Knicks play the Bulls.

_____

Application

14. Make your own geometric design on a piece of white paper. Color about 55% red, 20% blue, 15% green, and leave the remainder white.

MIXED APPLICATIONS

A carpenter is planning a cabinet 8 ft long with four doors. There are $1\frac{1}{2}$ in. stiles to be placed between the doors and at each end. How many inches wide must each door be?

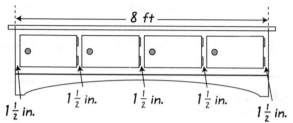

To find the width of each door:

a. Multiply to find the width of the 5 stiles.

$$5 \times 1\frac{1}{2} = \frac{5}{1} \times \frac{3}{2} = \frac{15}{2} = 7\frac{1}{2}$$

b. Subtract the width of the stiles from the total width. (Write measurements in the same units:

8 ft $\longrightarrow$ 8 × 12 in. = 96 in.)

$$\begin{array}{r} 96 \;\;= 95\frac{2}{2} \\ -\; 7\frac{1}{2} = \;\; 7\frac{1}{2} \\ \hline 88\frac{1}{2}, \text{ or } 88\frac{1}{2} \text{ in.} \end{array}$$

c. Divide by 4 to find the width of one door.

$$88\frac{1}{2} \div 4 = \frac{177}{2} \div \frac{4}{1}$$

$$= \frac{177}{2} \times \frac{1}{4} = \frac{177}{8} = 22\frac{1}{8}$$

Each door of the cabinet must be $22\frac{1}{8}$ in. wide.

Reminder

To subtract a mixed number from a whole number, you must rename the whole number as a mixed number.

Reminder

To divide fractions, multiply by the reciprocal of the divisor.

Guided Practice

1. One serving of lamb curry needs $3\frac{1}{2}$ oz. of meat. If Ved is cooking for 14 people, how many ounces of meat will he need? How many pounds?

$$14 \times 3\frac{1}{2} = \frac{14}{?} \times \frac{?}{2} = \frac{?}{2} =$$

_____ oz.

16 ounces equals 1 pound

_____ oz ÷ 16 = _____ lb, or

_____ lb, _____ oz.

2. On Monday, the stock of Ohio Edison opened at $19\frac{7}{8}$ and closed at $19\frac{1}{2}$. How much did it drop? If the stock drops the same amount each day for the next three days, what will be the final price?

3. One room in the community college is set aside as a computer laboratory. $\frac{2}{5}$ of the computer terminals are used for word processing. Of these, $\frac{3}{4}$ have color screens. What part of the total computers in the laboratory have color screens?

4. To mix a batch of muffins, Ezell will sift all of the dry ingredients into a bowl that holds 12 cups. Will $8\frac{1}{2}$ cups of flour, $2\frac{3}{4}$ cups of bran, and $\frac{1}{4}$ cup of baking powder all fit into the bowl? Explain.

5. An airline mechanic cuts a piece of tubing $9\frac{5}{8}$ in. long from a piece that is $33\frac{1}{2}$ in. long. The process of cutting removes an additional $\frac{1}{16}$ in. How much is left?

6. A carpenter cut an 8-ft board into 16-in. boards. How many boards did the carpenter make?

7. The circle graph shows the results of a survey taken to find students' preferred forms of exercise. Write a fraction in simplest form for each exercise activity selected in the student survey.

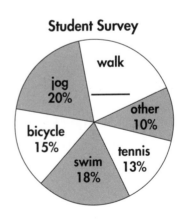

Student Survey

Find the prime factors of each number.

1. 30

2. 64

3. 189

_____   _____   _____

Find the least common multiple (LCM) of each set of numbers.

4. 6, 8

5. 13, 52

6. 10, 15, 20

Write a fraction or mixed number for the shaded regions.

7.

8.

_____   _____

Complete to rename the fraction in higher terms or lower terms.

9. $\dfrac{15}{40} = \dfrac{?}{8}$ _____

10. $\dfrac{12}{16} = \dfrac{?}{48}$ _____

11. $\dfrac{23}{25} = \dfrac{92}{?}$ _____

Rename each fraction in simplest form or lowest terms.

12. $\dfrac{8}{48}$ _____

13. $\dfrac{35}{55}$ _____

14. $\dfrac{42}{91}$ _____

Rename each pair of fractions so they have the lowest common denominator (LCD) as their denominators.

15. $\dfrac{5}{8}$ _____

16. $\dfrac{9}{10}$ _____

17. $\dfrac{11}{15}$ _____

$\dfrac{1}{3}$ _____ $\dfrac{5}{6}$ _____ $\dfrac{4}{45}$ _____

18. A friend says "I am confused by the terms greatest common factor and least common multiple." Write an explanation to help your friend better understand these terms.

_____

_____

_____

Rename each fraction or mixed number as a decimal. If necessary, round the answer to the nearest hundredth.

1. $\frac{7}{25}$ _____

2. $\frac{5}{9}$ _____

3. $38\frac{5}{6}$ _____

Rename each decimal as a fraction or mixed number in simplest form.

4. 0.45 _____

5. 0.64 _____

6. 12.08 _____

Write in order from least to greatest.

7. $\frac{2}{3}, \frac{3}{4}, \frac{3}{5}$

8. $\frac{11}{24}, \frac{5}{8}, \frac{3}{4}$

9. $\frac{5}{12}, \frac{7}{15}, \frac{3}{8}$

_____   _____   _____

Express each improper fraction as a whole number or a mixed number in simplest form.

10. $\frac{24}{15}$ _____

11. $\frac{91}{7}$ _____

12. $\frac{112}{12}$ _____

Express each mixed number as an improper fraction.

13. $6\frac{1}{6}$ _____

14. $3\frac{11}{12}$ _____

15. $10\frac{5}{9}$ _____

Write in order from greatest to least.

16. $7\frac{2}{3}, 7\frac{5}{8}, 7\frac{3}{4}$

17. $1\frac{11}{15}, 1\frac{3}{5}, 1\frac{2}{3}$

_____   _____

18. $\frac{42}{14}, 2\frac{13}{14}, 2\frac{6}{7}$

19. $\frac{216}{696}, \frac{16}{50}, .318$

_____   _____

20. Tito turned 18 years old today. Write mixed numbers in simplest form to show what his age will be in 8 months, 15 months, and 50 months.

_____

_____

Decide if the fractions are close to 0, $\frac{1}{2}$, or 1. Then estimate the sum or difference.

1. $\frac{1}{7} + \frac{9}{16}$ _____

2. $\frac{16}{18} - \frac{10}{12}$ _____

3. $\frac{8}{9} + \frac{2}{11} + \frac{9}{19}$ _____

Estimate the sum or difference.

4. $12\frac{1}{10}$
$+ 19\frac{2}{5}$
_____

5. $9\frac{7}{8}$
$- 2\frac{7}{20}$
_____

6. $17\frac{5}{12}$
$+ 18\frac{9}{18}$
_____

Add. Write each answer in simplest form.

7. $\frac{11}{18} + \frac{10}{18}$ _____

8. $\frac{2}{25} + \frac{23}{25}$ _____

9. $\frac{5}{8} + \frac{7}{8} + \frac{8}{8}$ _____

10. $\frac{5}{6} + \frac{9}{10}$ _____

11. $\frac{4}{7} + \frac{17}{42}$ _____

12. $\frac{7}{9} + \frac{3}{5}$ _____

13. $5\frac{4}{9}$
$+ 7\frac{2}{9}$
_____

14. $10\frac{3}{20}$
$+ 1\frac{3}{4}$
_____

15. $5\frac{7}{12}$
$+ \frac{7}{12}$
_____

16. Tyrone and Luciano are replacing a door. The doorway measures 7 ft $5\frac{3}{8}$ in. by $34\frac{1}{2}$ in. The door they install must be narrower than each side of the doorway by $\frac{5}{16}$ in. What are the dimensions of the new door? What is the perimeter of the new door?

_____

_____

17. Determine what number is added to get the sequence. Then find the next three numbers of the sequence.

a. $\frac{1}{2}, \frac{5}{6}, 1\frac{1}{6},$ _____, _____, _____

b. $\frac{3}{8}, 1\frac{1}{2}, 2\frac{5}{8},$ _____, _____, _____

Subtract. Write each answer in simplest form.

1. $\dfrac{8}{9} - \dfrac{2}{9}$ _____

2. $\dfrac{11}{24} - \dfrac{1}{24}$ _____

3. $\dfrac{41}{35} - \dfrac{16}{35}$ _____

4.
$$\begin{array}{r} \dfrac{11}{15} \\[2mm] -\ \dfrac{2}{5} \\ \hline \end{array}$$

5.
$$\begin{array}{r} \dfrac{3}{4} \\[2mm] -\ \dfrac{3}{7} \\ \hline \end{array}$$

6.
$$\begin{array}{r} \dfrac{9}{10} \\[2mm] -\ \dfrac{2}{8} \\ \hline \end{array}$$

7.
$$\begin{array}{r} 9\dfrac{11}{12} \\[2mm] -\ 3\dfrac{3}{12} \\ \hline \end{array}$$

8.
$$\begin{array}{r} 13\dfrac{1}{2} \\[2mm] -\ 2\dfrac{5}{22} \\ \hline \end{array}$$

9.
$$\begin{array}{r} 22\dfrac{8}{9} \\[2mm] -\ 9\dfrac{5}{7} \\ \hline \end{array}$$

10.
$$\begin{array}{r} 8\dfrac{1}{6} \\[2mm] -\ 2\dfrac{3}{4} \\ \hline \end{array}$$

11.
$$\begin{array}{r} 11 \\[2mm] -\ 5\dfrac{3}{9} \\ \hline \end{array}$$

12.
$$\begin{array}{r} 30\dfrac{4}{15} \\[2mm] -\ 9\dfrac{7}{45} \\ \hline \end{array}$$

Add or subtract as indicated.

13. $\dfrac{1}{4} + \dfrac{1}{2} - \dfrac{1}{3}$ _____

14. $\dfrac{5}{6} - \dfrac{3}{10} + \dfrac{1}{3}$ _____

15. $19\dfrac{1}{2} - 8\dfrac{1}{7} - 7\dfrac{1}{4}$ _____

16. $22 - 1\dfrac{5}{8} + 10\dfrac{1}{4}$ _____

17. Anka is setting her VCR to tape a sports program that begins at 5:30 P.M. The program could run anywhere from $2\dfrac{3}{4}$ to $3\dfrac{1}{4}$ hours. Should she stop taping before or after 9:00 P.M.? Explain.

_____

_____

_____

18. How is regrouping when adding mixed numbers different from regrouping when subtracting mixed numbers?

_____

_____

_____

Multiply.

1. $\frac{8}{9} \times \frac{27}{64}$ _____

2. $\frac{5}{9} \times \frac{10}{11}$ _____

3. $\frac{3}{7}$ of $49 _____

4. $2\frac{1}{4} \times 5\frac{1}{3}$ _____

5. $3\frac{1}{3} \times 3\frac{1}{5}$ _____

6. $1\frac{1}{18} \times 9$ _____

Divide.

7. $\frac{5}{8} \div \frac{3}{4}$ _____

8. $\frac{14}{15} \div \frac{7}{12}$ _____

9. $8 \div \frac{6}{5}$ _____

10. $1\frac{1}{5} \div 2\frac{1}{7}$ _____

11. $4\frac{1}{2} \div 2\frac{2}{3}$ _____

12. $6\frac{2}{3} \div 9$ _____

 Estimate each product or quotient. Then use a calculator to check your estimates.

13. $4\frac{1}{3} \times 29\frac{1}{2}$

14. $17\frac{3}{5} \div 9\frac{1}{4}$

15. $2\frac{7}{8} \div 15$

_____ _____ _____

Assume that you can jump 3 ft on Earth. The table shows the heights of the same jump on some other planets and on the Moon.

Planet	Height Jumped
Jupiter	1 ft $3\frac{1}{2}$ in.
Saturn	2 ft $7\frac{1}{4}$ in.
Earth	3 ft
Mercury	8 ft $1\frac{1}{4}$ in.
Moon	18 ft 9 in.

16. Find the total distance, in inches, of four 3-ft Earth jumps on Saturn. Hint: Change 2 ft $7\frac{1}{4}$ in. to inches.

_____

17. The jump on the moon is how many times the same jump on Saturn?

_____

 18. Is the jump on Mercury more than $6\frac{1}{4}$ times the same jump on Jupiter? Explain.

_____

23-25 CUMULATIVE REVIEW

Rename each fraction as a percent.

1. $\dfrac{52}{100}$ _____

2. $\dfrac{9}{10}$ _____

3. $\dfrac{3}{25}$ _____

Use a calculator to rename each fraction as a percent.

4. $\dfrac{11}{15}$ _____

5. $\dfrac{3}{16}$ _____

6. $\dfrac{7}{12}$ _____

Rename each percent as a fraction in simplest form.

7. 80% _____

8. 70% _____

9. 15% _____

10. 6% _____

11. 84% _____

12. 71% _____

13. A long-playing record makes $33\frac{1}{3}$ revolutions per minute. After 500 revolutions, how long has it been playing?

_____

_____

14. The total height of a cabinet, including its top and base, is 30 inches. Its base is 4 inches and it has a $1\frac{1}{2}$-in. thick top. Four equal-size drawers fit in the remaining space, with $\frac{3}{4}$ inch between each pair of drawers. What is the height of each drawer?

_____

_____

_____

15. A $6\frac{1}{2}$-meter pole is set $2\frac{3}{4}$ meters in the ground. How much of the pole is above ground? What fractional part of the pole is above ground? (Hint: the fractional part is the length of the pole above the ground divided by the total length of the pole.) What percent of the pole is above the ground?

_____

_____

_____

_____

_____

ANSWER KEY

LESSON 1 (pages 2–3)
 1. 1, 2, 3, 4, 6, 8, 12, 24
 3. 1, 2, 4, 5, 10, 20
 5. 1, 2, 3, 6, 9, 18
 7. 1, 2, 3, 4, 5, 6, 10, 12, 15, 20, 30, 60
 9. $2 \times 2 \times 2 \times 2 \times 2$
 11. $2 \times 3 \times 11$ **13.** 5×19
 15. Answers may vary.
  Start with 2 and divide by prime numbers.
  Continue to divide until you get all the
  prime factors.

LESSON 2 (pages 4–5)
 1. a. 2, 4, 6, 8, 10, 12, 14, 16, 18, 20;
  3, 6, 9, 12, 15, 18, 21, 24, 27, 30
  b. 6, 12, 18
 3. 2, 4, 6, 8, 10, 12, 14, 16, 18, 20
 5. 6, 12, 18, 24, 30, 36, 42, 48, 54, 60
 7. 20, 40, 60, 80, 100, 120, 140, 160, 180, 200
 9. 6 **11.** 20 **13.** 24
 15. Col. A: 6, 20, 42, 72;
  Col. B: 14, 12, 10, 21;
  Col. C: 6, 35, 143, 323

LESSON 3 (pages 6–7)
 1. b. $\frac{7}{20}$ **3.** $\frac{3}{6}$ **5.** $\frac{7}{7}$ **7.** $4\frac{2}{3}$
 9. $3\frac{4}{12}$ **11.** Answers may vary.

LESSON 4 (pages 8–9)
 1. a. $\frac{18}{24}$ **b.** $\frac{3}{4}; \frac{6 \div 2}{8 \div 2} = \frac{3}{4}$
 3. 6 **5.** 4 **7.** 24
 9. $\frac{1}{3}; \frac{9 \div 9}{27 \div 9} = \frac{1}{3}$
 11. $\frac{5}{8}; \frac{20 \div 4}{32 \div 4} = \frac{5}{8}$
 13. $\frac{3}{14}; \frac{21 \div 7}{98 \div 7} = \frac{3}{14}$
 15. $420 + 150 + 120 = 690$ kg;
  Cement: $\frac{420}{690} = \frac{420 \div 30}{690 \div 30} = \frac{14}{23}$,
  Stone: $\frac{150}{690} = \frac{5}{23}$,
  Sand: $\frac{120}{690} = \frac{4}{23}$
 17. $\frac{9}{15}$

LESSON 5 (pages 10–11)
 1. a. 6, 12, 18, 24, 30, 36;
  5, 10, 15, 20, 25, 30; LCM = 30; LCD = 30
  b. $\frac{25}{30}; \frac{5 \times 5}{6 \times 5} = \frac{25}{30}$,
  $\frac{24}{30}; \frac{4 \times 6}{5 \times 6} = \frac{24}{30}$
 3. 20 **5.** 8 **7.** 30 **9.** 24 **11.** $\frac{4}{16}, \frac{7}{16}$
 13. $\frac{2}{12}, \frac{9}{12}$ **15.** $\frac{15}{20}, \frac{18}{20}$ **17.** $\frac{6}{20}, \frac{3}{20}$ **19.** $\frac{36}{60}, \frac{35}{60}$

LESSON 6 (pages 12–13)
 1. a. 0.625; 0.625
  b. $2 \div 3 = 0.666\ldots$; 4.7
 3. d **5.** 0.75 **7.** 7.8 **9.** 0.6875
 11. $0.8333\ldots = 0.83$

13. $18.6666\ldots = 18.67$
15. $7.1111\ldots = 7.11$
17. 0.48
19. 0.15; $\frac{3 \times 5}{5 \times 20} = \frac{15}{100} = 0.15$
21. 80.88; $80\frac{22 \times 4}{25 \times 4} = 80\frac{88}{100} = 80.88$

LESSON 7 (pages 14–15)
 1. $0.025 = \frac{25}{1,000} = \frac{25 \div 25}{1,000 \div 25} = \frac{1}{40}$
 3. 31 **5.** $\frac{8}{10} = \frac{4}{5}$ **7.** $1\frac{23}{100}$
 9. $\frac{75}{100} = \frac{3}{4}$ **11.** $\frac{2}{10} = \frac{1}{5}$
 13. $\frac{8}{1,000} = \frac{1}{125}$ **15.** $5\frac{4}{10} = 5\frac{2}{5}$
 17. $9\frac{77}{100}$ **19.** $56\frac{50}{1,000} = 56\frac{1}{20}$
 21–26. Answers may vary.
 21. $\frac{1}{2}$; $0.50 = \frac{50}{100} = \frac{1}{2}$
 23. $\frac{21}{50}$; $0.42 = \frac{42}{100} = \frac{21}{50}$ **25.** $0.31 = \frac{31}{100}$

LESSON 8 (pages 16–17)
 1. > **3.** < **5.** < **7.** >
 9. $\frac{1}{6}, \frac{1}{2}, \frac{2}{3}$ **11.** $\frac{11}{18}, \frac{2}{3}, \frac{7}{9}$ **13.** $\frac{3}{4}, \frac{11}{16}, \frac{5}{8}$
 15. $(\frac{1}{2}) < (1)$ **17.** $(\frac{1}{2}) < (1)$

LESSON 9 (pages 18–19)
 1. a. $3\frac{2}{3}$ **b.** $\frac{51}{8}$ **c.** Arrows pointing to $3\frac{2}{3}$ and $6\frac{3}{8}$
 3. $1\frac{6}{8} = 1\frac{3}{4}$ **5.** 3 **7.** $9\frac{2}{10} = 9\frac{1}{5}$ **9.** $\frac{10}{9}$
 11. $\frac{40}{3}$ **13.** $\frac{35}{12}$
 15. Arrows pointing to $\frac{23}{4}$ and $\frac{32}{4}$
 17. 23 quarter hours; $5\frac{3}{4} = \frac{23}{4}$

LESSON 10 (pages 20–21)
 1. $8\frac{1}{3} = 8\frac{2}{6}, 8\frac{1}{2} = 8\frac{3}{6}$; Order: $8\frac{1}{3}, 8\frac{1}{2}, 8\frac{5}{6}$
 3. $1\frac{5}{6}, 1\frac{7}{9}, 1\frac{2}{3}$ **5.** $9\frac{3}{4}, 9\frac{2}{3}, 8\frac{4}{5}$
 6. $7\frac{1}{8}, 7\frac{1}{9}, 7\frac{1}{10}$ **7.** $1\frac{4}{11}, \frac{14}{11}, 1\frac{2}{11}$
 9. $4\frac{3}{8}$ lb is heaviest. $4\frac{3}{8} = 4\frac{6}{16}$; $4\frac{1}{4} = 4\frac{4}{16}$

LESSON 11 (pages 22–23)
 1. 1; $1 + 0 = 1$ **3.** 13; $7 + 6 = 13$
 5. $\frac{1}{2}$; $0 + \frac{1}{2} = \frac{1}{2}$ **7.** 1; $0 + 1 = 1$
 9. $\frac{1}{2}$; $\frac{1}{2} - 0 = \frac{1}{2}$ **11.** $1\frac{1}{2}$; $1 + 0 + \frac{1}{2} = 1\frac{1}{2}$
 13. 15; $9 + 6 = 15$ **15.** $14\frac{1}{2}$; $15\frac{1}{2} - 1 = 14\frac{1}{2}$
 17. Answers may vary.
  a. $2\frac{1}{9}$ **b.** $4\frac{1}{15}$ **c.** $5\frac{1}{7}$ **d.** $4\frac{1}{7}$

LESSON 12 (pages 24–25)
 1. $1\frac{2}{6}$ **3.** $\frac{3}{8}; \frac{6}{16} = \frac{3}{8}$ **5.** $1\frac{2}{5}; \frac{7}{5} = 1\frac{2}{5}$
 7. $1\frac{5}{6}; \frac{22}{12} = 1\frac{10}{12} = 1\frac{5}{6}$ **9.** $\frac{2}{3}; \frac{10}{15} = \frac{2}{3}$
 11. $\frac{1}{2}; \frac{18}{36} = \frac{1}{2}$ **13.** $2\frac{1}{4}; \frac{27}{12} = 2\frac{3}{12} = 2\frac{1}{4}$
 15. $2; \frac{20}{10} = 2$ **17.** D **19.** $\frac{23}{16} = 1\frac{7}{16}; \frac{15}{16} + \frac{8}{16} = \frac{23}{16}$

LESSON 13 (pages 26–27)
 1. $\frac{1}{5} = \frac{1 \times 4}{5 \times 4} = \frac{4}{20}$ **3.** $\frac{7}{18}; \frac{4}{18} + \frac{3}{18} = \frac{7}{18}$
 5. $\frac{2}{3}; \frac{9}{15} + \frac{1}{15} = \frac{10}{15} = \frac{2}{3}$ **7.** $1\frac{1}{2}; \frac{12}{16} + \frac{12}{16} = \frac{24}{16} = 1\frac{8}{16} = 1\frac{1}{2}$
 9. $1\frac{1}{49}; \frac{28}{49} + \frac{22}{49} = \frac{50}{49} = 1\frac{1}{49}$ **11.** $\frac{13}{20}; \frac{3}{20} + \frac{4}{20} + \frac{6}{20} = \frac{13}{20}$
 13. Tuesday $\frac{1}{6} + \frac{1}{3} = \frac{1}{6} + \frac{2}{6} = \frac{3}{6} = \frac{1}{2}$
 15. Moniesha, $\frac{139}{120} > \frac{137}{120}$

LESSON 14 (pages 28–29)
1. $16\frac{6}{8} = 16\frac{3}{4}$ **3.** $5\frac{1}{4} = 5\frac{2}{8}; 8\frac{5}{8}$
5. $10\frac{1}{6} = 10\frac{2}{12}, 11\frac{1}{4} = 11\frac{3}{12}, 21\frac{5}{12}$
7. $5\frac{8}{15}$ **9.** $13\frac{4}{9}; 13\frac{8}{18} = 13\frac{4}{9}$
11. $34\frac{14}{15}; 34\frac{28}{30} = 34\frac{14}{15}$ **13.** $36\frac{7}{12}$
15. $9\frac{1}{3}$ in. by 6 in. frame is greater.;
  $6\frac{1}{2} + 6\frac{1}{2} + 8\frac{3}{4} + 8\frac{3}{4} = 30\frac{1}{2};$
  $9\frac{1}{3} + 9\frac{1}{3} + 6 + 6 = 30\frac{2}{3}$
17. $11\frac{1}{3}$

LESSON 15 (pages 30–31)
1. $\frac{7}{16}$; about $\frac{1}{2}$ **3.** $\frac{3}{5}; \frac{9}{15} = \frac{3}{5}$
5. $\frac{1}{6}; \frac{2}{12} = \frac{1}{6}$ **7.** $1; \frac{12}{12} = 1$
9. $\frac{2}{3}; \frac{14}{21} = \frac{2}{3}$
11. $\frac{11}{16}; \frac{15}{16} + \frac{3}{16} = \frac{18}{16}; \frac{18}{16} - \frac{7}{16} = \frac{11}{16}$
13. $1; \frac{8}{8} = 1$
15. $\frac{20}{50} - \frac{16}{50} = \frac{4}{50} = \frac{2}{25}$

LESSON 16 (pages 32–33)
1. $\frac{4}{5} = \frac{4 \times 3}{5 \times 3} = \frac{12}{15} - \frac{1}{15} = \frac{11}{15}$
3. $\frac{1}{6}$ **5.** $\frac{4}{15}$
7. $\frac{11}{15}$ **9.** $\frac{1}{4}$
11. $\frac{7}{12}$ **13.** $\frac{71}{72}$ **15.** $\frac{7}{12}$
17. 30 in.; 30 in. $= \frac{5}{6}$ yd, $1\frac{2}{3}$ yd $= \frac{5}{3}$ yd; $\frac{5}{3} - \frac{5}{6} = \frac{5}{6}$

LESSON 17 (pages 34–35)
1. $4\frac{2}{5}$ **3.** $7\frac{2}{15}$
5. $7\frac{1}{2}; 7\frac{2}{4} = 7\frac{1}{2}$ **7.** $14\frac{2}{9}$
9. $8\frac{7}{15}$ **11.** $4\frac{7}{90}$ **13.** 0
15. $2\frac{3}{4} + 1\frac{1}{2} = 3\frac{5}{4} = 4\frac{1}{4}; 4\frac{1}{2} - 4\frac{1}{4} = \frac{1}{4}$; Yes, Cleon has enough flour and $\frac{1}{4}$ cup more.

LESSON 18 (pages 36–37)
1. $6 + \frac{5}{5} + \frac{2}{5} = 6\frac{7}{5}$
3. $8\frac{21}{12} - 4\frac{10}{12} = 4\frac{11}{12}$
5. $2\frac{2}{3}$ **7.** $6\frac{1}{2}; 6\frac{3}{6} = 6\frac{1}{2}$
9. $7\frac{1}{2}$ **11.** $3\frac{9}{20}$ **13.** $8\frac{33}{40}$

LESSON 19 (pages 38–39)
1. $\frac{3 \times 7}{5 \times 8} = \frac{21}{40}$ **3.** $\frac{1}{4}$
5. $\frac{14}{27}$ **7.** 10
9. $18 **11.** $2.40 **13.** $\frac{1}{24}$
15. $\frac{1}{10}$ Chinese; $\frac{9}{10}$ not Chinese

LESSON 20 (pages 40–41)
1. $\frac{25}{4} \times \frac{12}{5} = \frac{5 \times 3}{1 \times 1} = \frac{15}{1} = 15$
3. 2 **5.** 32
7. $5\frac{1}{4}; \frac{21}{4} = 5\frac{1}{4}$
9. $2\frac{1}{2}; \frac{5}{2} = 2\frac{1}{2}$
11. $6\frac{1}{4}; \frac{25}{4} = 6\frac{1}{4}$
13. $3\frac{3}{4}; \frac{15}{4} = 3\frac{3}{4}$
15. > **17.** <
19. 1000

LESSON 21 (pages 42-43)
1. $\frac{6}{7} \times \frac{4}{3} = \frac{8}{7} = 1\frac{1}{7}$
3. 3 **5.** $\frac{8}{9}$
7. 8 **9.** $\frac{1}{24}$ **11.** $\frac{1}{12}$
13. 24 bowls

LESSON 22 (pages 44–45)
1. $\frac{9}{4} \times \frac{4}{5} = \frac{9}{5} = 1\frac{4}{5}$
3. 2 **5.** $\frac{9}{28}$
7. $1\frac{7}{17}$ **9.** $4\frac{4}{5}$ **11.** $\frac{4}{5}$
13. 4 **15.** $\frac{1}{3}$ **17.** $6\frac{3}{4}$
19. $16\frac{1}{2} \div 1\frac{1}{3} = 12\frac{3}{8}$; 12 shirts

LESSON 23 (pages 46–47)
1. 15%
3. $5 \div 8 = 62.5$; 62.5%
5. 8% **7.** 40%
9. 30% **11.** 4%
13. 87.5% **15.** 60%
17. about 66.67%
19. Advertising Day:
  advertising: $\frac{72}{120} = 60\%$;
  regular customer: $\frac{24}{120} = 20\%$;
  friend recommends: $\frac{6}{120} = 5\%$;
  driving by: $\frac{18}{120} = 15\%$
  Normal Day:
  advertising: $\frac{20}{90} =$ about 22.22%;
  regular customer: $\frac{55}{90} =$ about 61.11%;
  friend recommends: $\frac{9}{90} = 10\%$;
  driving by: $\frac{6}{90} =$ about 6.67%

LESSON 24 (pages 48–49)
1. a. $44\% = \frac{44}{100}$
  b. $\frac{44}{100} = \frac{44 \div 4}{100 \div 4} = \frac{11}{25}$
3. $\frac{9}{10}$
5. $\frac{7}{20}$
7. $\frac{2}{25}$
9. $\frac{27}{100}$
11. 72%; 27%
13. Rename 64% as $\frac{16}{25}$.

LESSON 25 (pages 50–51)
1. $\frac{14}{1} \times \frac{7}{2} = \frac{7}{1} \times \frac{7}{1} = 49$; 49 oz = 3 lb 1 oz
3. $\frac{3}{4} \times \frac{2}{5} = \frac{3}{10}; \frac{3}{10}$ have color screens.
5. $23\frac{13}{16}$ in.
7. walk: $\frac{6}{25}$; jog: $\frac{1}{5}$; swim: $\frac{9}{50}$; bicycle: $\frac{3}{20}$; tennis: $\frac{13}{100}$; other: $\frac{1}{10}$

CUMULATIVE REVIEW: LESSONS 1–5 (page 52)
1. $2 \times 3 \times 5$
3. $3 \times 3 \times 3 \times 7$
5. 52 **7.** $\frac{12}{15}$ or $\frac{4}{5}$ **9.** 3 **11.** 100
13. $\frac{7}{11}$ **15.** $\frac{15}{24}, \frac{8}{24}$ **17.** $\frac{33}{45}, \frac{4}{45}$

CUMULATIVE REVIEW: LESSONS 6–10 (page 53)

1. 0.28 **3.** 38.83

5. $\frac{16}{25}$ **7.** $\frac{3}{5}, \frac{2}{3}, \frac{3}{4}$ **9.** $\frac{3}{8}, \frac{5}{12}, \frac{7}{15}$

11. 13 **13.** $\frac{37}{6}$

15. $\frac{95}{9}$ **17.** $1\frac{11}{15}, 1\frac{2}{3}, 1\frac{3}{5}$

19. $\frac{16}{50}, .318, \frac{216}{696}$

CUMULATIVE REVIEW: LESSONS 11–14 (page 54)

1. $\frac{1}{2}$ **3.** $1\frac{1}{2}$

5. $10 - 2 = 8$

7. $1\frac{1}{6}$ **9.** $2\frac{1}{2}$ **11.** $\frac{41}{42}$

13. $12\frac{2}{3}$ **15.** $6\frac{1}{6}$

17. a. $\frac{1}{3}; 1\frac{1}{2}, 1\frac{5}{6}, 2\frac{1}{6}$ **b.** $1\frac{1}{8}; 3\frac{3}{4}, 4\frac{7}{8}, 6$

CUMULATIVE REVIEW: LESSONS 15–18 (page 55)

1. $\frac{2}{3}$ **3.** $\frac{5}{7}$ **5.** $\frac{9}{28}$ **7.** $6\frac{2}{3}$ **9.** $13\frac{11}{63}$

11. $5\frac{2}{3}$ **13.** $\frac{5}{12}$ **15.** $4\frac{3}{28}$

17. Before 9:00 P.M.; $9 - 5\frac{1}{2} = 3\frac{1}{2}$ hours; $3\frac{1}{2} > 3\frac{1}{4}$, so $3\frac{1}{2}$ hours is more time than is needed.

CUMULATIVE REVIEW: LESSONS 19–22 (page 56)

1. $\frac{3}{8}$ **3.** \$21 **5.** $\frac{32}{3} = 10\frac{2}{3}$

7. $\frac{5}{6}$ **9.** $\frac{20}{3}$, or $6\frac{2}{3}$ **11.** $1\frac{11}{16}$

13. $4 \times 30 = 120$; 127.83

15. $3 \div 15 = \frac{1}{5}$; 0.19

17. $7\frac{1}{5}$ times

CUMULATIVE REVIEW: LESSONS 23–25 (page 57)

1. 52% **3.** $\frac{12}{100} = 12\%$

5. 18.75% **7.** $\frac{4}{5}$

9. $\frac{3}{20}$ **11.** $\frac{21}{25}$

13. 15 minutes

15. $3\frac{3}{4}$ m above ground; $\frac{15}{26}$ of pole above ground; $\frac{15}{26}$ = about 57.7%